IN THE
FOOTSTEPS
OF EVE

IN THE FOOTSTEPS OF EVE

THE MYSTERY OF HUMAN ORIGINS

LEE R. BERGER, PH.D.

WITH

BRETT HILTON-BARBER

ADVENTURE PRESS

NATIONAL GEOGRAPHIC
WASHINGTON, D. C.

Library of Congress Cataloging-in-Publication Data

Berger, Lee R.
    In the footsteps of Eve : exploring the mysteries of human origins / Lee R. Berger with
Brett Hilton-Barber.
        p. cm.
    ISBN 0-7922-7682-5
    1. Human beings--Origin. I. Hilton-Barber, Brett. II. Title.

GN281 .B47 2000
599.93'8—dc21

                                                                    00-028071
                                                                        CIP

Interior design by Carol Norton.
Printed in U.S.A.

To Jackie, Megan, and Matthew

*To Josie and James, who have brought joy and inspiration
to the continuation of the species*

CONTENTS

INTRODUCTION 1

1. FIRST STEPS 16

2. AFRICA CALLS 45

3. THE UNDISCOVERED COUNTRY 68

4. EAST SIDE STORY 101

5. FOSSIL FORENSICS 122

6. SLAUGHTERING SACRED COWS 140

7. LONG ARMS—SHORT LEGS 164

8. KNEE-JERK REACTION 188

9. HOMINIDS ALIVE 212

10. SKELETONS IN THE CLOSET 226

11. THE ROBUST ENIGMA 258

12. FOOTSTEPS OF EVE 276

13. FINAL STEPS 302

INDEX 312

BIBLIOGRAPHY AND RELATED READINGS 320

ACKNOWLEDGMENTS 324

INTRODUCTION

〜

The cold winter rain that had lashed the coastline for almost two days finally let up, and slivers of late afternoon sunlight emerged from behind heavy clouds. The wind died down and with it went the white seafroth capping the gray-green swells of the southern Atlantic. Kelp gulls spiraled in salty gusts of air that tailed a cold front passing above the southern African coastline, pushing the rain inland over the desolate landscape. It was clear the weather was changing for the better. This was a welcome respite for the inhabitants around the lagoon, though they had been protected from the storm's full force by a mile-wide spit of duneland jutting into the ocean. The sea side of this modest peninsula continued to endure the foaming fury of the restless southern current, but as the wind died down in the lagoon, choppy waves subsided into ripples. The tide was high, and the seawater lapped onto sand just a few yards from the shrubs at the base of the surrounding dunes.

Slowly, the dune brush came alive as tiny animals emerged from their rain-sodden lairs, stretching tired, cramped muscles to generate warmth. One of the first to ease out of the undergrowth was a steenbok, leaving one-inch prints in the wet sand as it minced its way onto the narrow strip of beach. The little gray-brown buck, raising its button-size black nose as high as its neck would allow, cautiously sniffed the air for the scent of danger. To pick up even the faintest sound of hostility, it flicked its ears back and forth. The wind and the crash of waves on the other side of the dunes made hearing difficult. As one of Africa's smallest antelopes, it had to be wary of every predator, from the lowly jackal to the mighty lion. Understandably, two days of unrelenting rain had sharpened the pangs of hunger among all the animals, including a large male hyena that had emerged about half a mile away from its burrow in the dune field. Motivated by an empty stomach, the brown hyena had come out to search for food in the daylight. It headed first for the ocean-side beach, but something made it switch direction toward the calmer waters of the lagoon, where a succulent morsel may have washed ashore.

The steenbok froze. It spotted the hyena emerging from a bush several hundred yards away. Unlike bigger antelopes, which would have bounded away, the steenbok's survival lay in its ability to fuse stillness and silence. It skipped into the scrub and stood absolutely motionless, its color acting as a surprisingly good camouflage against the dull green tones of the vegetation. The hyena padded along the narrow beach, leaving large doglike tracks along the high-tide mark as it meandered in an easterly direction. It passed within just three yards of the steenbok. Ten minutes after the hyena had moved on, the little buck stepped out and began nibbling at the foliage on the edge of the sand. Another sound made it freeze once more. Something else was coming.

A young woman paused at the top of the dune and looked back toward the ocean. She shifted the dead seal pup into her left hand, exchanging its weight for the long digging stick and pieces of driftwood that she carried in her right. Behind her, a mile away, was the spot where she had found the seal, washed up among the seaweed and storm debris. She smiled. The seal was young and full of fat. It would warm her small tribe, an extended family group of 25 men, women, and children. She was pleased that she had gone out when she sensed the rain was starting to lift. Being first on the beach meant gathering the best pickings of anything edible that had washed up during the storm, something her mother's mother had taught her.

Standing just over five feet tall, she had reached her full height and was considered at 13 years old to be an adult. The yellow-brown skin on her face and arms was already wrinkled from the extreme heat and cold, which also shaped the harsh, treeless landscape. Her body was covered in soft springbok skins; despite being damp from the rain, they gave her a feeling of warmth and comfort. Her short, curly dark hair was permeated with red ocher dust, signifying her status as a single woman of childbearing age. Catching her breath from her trek over the dunes, she turned, looked over the lagoon, and momentarily contemplated her future. She knew she would have a child soon and hoped it would survive. The father would either be a man from her small tribe or from one of the nomadic neighboring groups three miles away across the water, on the northern side of the lagoon. She was not particularly attracted to any of the men in her group who had shown interest in her, and felt torn between her desire to stay close to her mother and the need for her own mate.

The wind picked up and the cold cut through her thoughts; she realized that she was standing at the top of an exposed dune. The rising force of the inshore wind had already dried the dune sand, though the rain had stopped just a short time ago. She glanced around. The territory was familiar to her, but she had learned to be watchful of changes in her environment. Her brother told her that a small pride of lions had been seen in the area only days ago. She started walking down the dune toward the lagoon, her bare feet stepping carefully along the narrow trail carved out by her tribe and the hooves and feet of countless animals looking for an easy way through the prickly plants.

As she reached the bottom of the dune, the sand beneath her feet became wetter and she had to step carefully. She grinned as she felt the soft sand ooze up between her toes, and she instinctively curled them in response to the ticklish sensation. Slipping slightly as she reached the base of the dune, she regained her balance and stepped onto the firmer sand along the strand. She paused again and looked around, admiring the way her footprints were so cleanly etched on the face of the dune. The wind was getting stronger and the sunlight had disappeared completely, indicating that a second cold front would soon hit the shoreline. She walked briskly toward the shelters of skin and bushwood that her people had erected in the fold of dunes a mile ahead. Her step quickened in anticipation of the fire that she knew they would be making with the wood kept dry over the past two days. Her mother would be delighted with the seal, and the men would make a fuss over her as she dried her skin clothing in front of the flames. She noticed two sets of tracks in front of her, instantly recognizing that a steenbok had walked on the beach and a hyena had passed. Her eyes followed the steenbok prints back into the bush. She knew it was close. She briefly thought of trying to flush it from its cover and of hitting it with her stick, but she already had the seal and harsh weather was closing in again. Looking ahead, she continued down to the edge of the lagoon.

Although the base of the dune remained wet, the crest had dried considerably from the blowing wind. Within a few minutes, the dry sand at the top began to collapse in sheets down the face of the dune. More and more dry sand cascaded down the slope, causing a small avalanche. A six-inch layer of the dune loosened and skidded down to the beach some 30 yards below. The girl looked back and saw the startled steenbok spring away. Although she was not to know it, the sand slide buried her footprints, the steenbok's and hyena's as well, beneath a layer of shell-rich sand.

Within half an hour the rain returned and the sand covering the prints was saturated, beginning a process that would preserve her prints in sandstone for millennia. Over a hundred thousand years later, her footprints would be found, and people she would never know would give her a name: Eve.

On an early summer morning 117,000 years later, I found myself driving toward Langebaan Lagoon along West Coast Road through the thick veil of fog that often shrouds South Africa's Atlantic seaboard. The onshore breeze, chilled overnight by the icy Benguela Current, sweeps down Africa's western coastline and reacts with thermodynamic drama as it meets the day's early heat rising from the coastal plains. Great moist mountains of mist sweep in over the waves breaking on the rocks and beaches, blowing across the scrubland before dissipating several miles inland. Driving along West Coast Road out of Cape Town in conditions such as this can be hazardous. One is struck by how suddenly the bright sunshine disappears and visibility is reduced to ghostly silhouettes in the wet half-light. Today was precisely such a day. I switched on my headlights as I passed through a wall of low clouds, tightening my grip on the steering wheel as I concentrated on the road ahead. I was on my way to meet Eve, and the last thing I needed was a collision with an oncoming vehicle.

I was en route to take another look at the astounding set of fossilized human footprints discovered by my colleague geologist Dave Roberts in August 1995 on a beach at Langebaan Lagoon, an hour and a half north of Cape Town. The survival of these prints, laid down in the wet sand countless years ago by someone walking over the dunes to the seashore, is remarkable in and of itself. What is quite extraordinary is that the hyena and steenbok tracks were also preserved, hence the inspiration to reconstruct the scenario in which they may have all passed by each other within the space of less than an hour. But these were neither the oldest footprints of hominids yet discovered nor are they the best preserved—the Laetoli hominid prints in Tanzania are more than 3.5 million years older. The excitement knotting my stomach had more to do with the fact that these foot tracks in the sand dated back to the emergence of the first *anatomically* modern humans, a point triumphantly echoed in the popular press, which, after Roberts and I announced the findings, dubbed the fossil impression left in sandstone, "Eve's footprints."

"Eve," of course, refers not to the biblical Eve but to a mitochondrial Eve, the proposed genetic progenitor of all living humans. The concept of a mitochondrial Eve is a scientific theory first coined by geneticists at the University of California during the 1980s. The term is used to illustrate the theory that all the world's people are descendants of a small population of anatomically modern humans that existed between 100,000 and 200,000 years ago. Their argument was based on studies in molecular biology, particularly the occurrence of mitochondrial DNA (mtDNA), genes that are inherited only through the female lineage. The California team devised a statistical model to measure the rate of mtDNA mutation over time. What they found was that different populations of humans living in diverse areas of the world exhibited different degrees of variation. The variation between Africans was found to be greater than that between any other population grouping, indicating that people from that continent are older than those from anywhere else. Their genes have had more time to mutate. The inescapable logic of this argument is that all modern humans are in fact descendant from a single female living in Africa over 100,000 years ago. More recent research in other genetic areas, including Y-chromosome studies that look at paternal inheritance, has added its weight to this argument, and the same conclusion has been reached: Our genes tell us that we all share a very recent African origin.

The broader conclusion of this research, which has been combined with fossil evidence, has become known as the Out of Africa theory. This holds that anatomically modern humans evolved in Africa and that fully modern populations of Africans eventually expanded into the rest of the world, replacing existing populations of older *Homo* species. There has been vociferous criticism of this argument and the validity of the methodology questioned, but support for an Out of Africa model is growing. More compelling than the genetic theory is the physical evidence in the form of fossils, which points to the recent African ancestry of all modern humans; even more compelling than this is the latest genetic research on the San peoples of southern Africa (often referred to as Bushmen): Combined with archaeological evidence, the research now strongly indicates that not only are all humans African in their origin, but our roots may well be in southern Africa—where the most consistent archaeological traces of the emergence of modern humans are to be found along the South African Cape coast.

After about half an hour on the road, the fog began thinning. As the rising sun burned the mist off the surrounding plains, I glanced out to my left toward the ocean over low land covered with dwarfed shrubs. This part of the coast, known as the Strandveld, is unremarkable to the casual eye, consisting of scrubland with an occasional granite outcrop breaking the horizon line. The vegetation—collectively called fynbos, or "fine bush" in Afrikaans—is stunted by a climate characterized by strong winds and relatively low winter rainfall. The apparent uniformity of fynbos is misleading. This little plant kingdom, confined to a narrow strip of land around the Cape of Good Hope coast, actually boasts the highest diversity of species per square yard of any plant community in the world. Despite its rich diversity, fynbos offers very little in the way of nutrition. Yet it still manages to support a surprising abundance of game, from poisonous snakes like the dreaded Cape cobra to large mammals like the eland. The beauty of fynbos is apparent only in the spring, when a profuse flowering transforms the gray-green and mustard scrub into an intricate biomass carpet of red, yellow, and purple. When the Dutch colonist Jan van Riebeeck, who established the first permanent European outpost in the Cape in 1652, first saw this area, he expressed the view that "there is no land in the whole world so barren and unblessed by the Lord God." But van Riebeeck clearly did not see the fynbos in spring, and was surely not a paleontologist. The attraction of the West Coast is not immediately apparent, and neither is its prehistorical significance, which is largely hidden in the dune middens and limestone sediments that constitute the coastline. Yet nowhere else in the world is there an area where an unbroken archaeological record spans the transition from primitive to modern human. From the 250,000-year-old Saldanah Man skull to the Koeberg nuclear power station one passes on the road to Langebaan, a comprehensive history of the modern human condition is to be found in this unassuming landscape.

Ironically, the same barren nature of the landscape that appalled van Riebeeck is one of the main clues as to why a primitive form of *Homo sapiens* adopted more modern morphological characteristics and took on the trappings of what is recognizable contemporary human behavior. Humanity is a product of Africa. We are what we are today because we've been shaped by our environment—and it was the African environment that hosted almost every major evolutionary change we've experienced on our journey toward being human. This includes the divergence of our lineage from that of an ancestral chimpanzee five to seven million years ago to the early experiments in bipedalism from about four million years ago; from the leap in brain size to the first stone tool technologies; from the emergence of our own genus some 2.5 million years ago to the conquest of fire over a million years ago and the morphological transition to the modern human form. These are all developments that have a "made in Africa" stamp on them.

It is no great mystery why this should be so. The size and diversity of Africa create conditions conducive to evolutionary change; add to this that there is more habitable landmass on this continent than on any other, and it is not surprising that an explosion of primate species in Africa ultimately led to the evolution of humankind. The great Miocene rain forests that blanketed the continent virtually from coast to coast over a 15-million-year period provided ample food and protection for primates, the perfect developmental hothouse. Even when these forests began retreating in the face of the great aridification accompanying the beginning of the Pliocene some five million years ago, Africa remained more habitable than the Northern Hemisphere continents, which froze under the grip of an ice age.

So if the continent had so much going for it, why would modern humans have emerged in a place as bleak as the Western Cape coastline? The possible answer is that they didn't have much choice. There is increasing evidence that a series of climate changes caused the expanse of desert conditions in the South African interior, areas known today as the Kalahari and the Karoo. This would have forced the dispersal of populations of archaic *Homo sapiens*, who occupied this landscape about 300,000 years ago. Those that moved southward and westward soon found themselves trapped by the Atlantic and Indian Oceans, the Namib Desert to the north, and the Kalahari and Karoo deserts they'd left behind. Over tens of thousands of years, these populations would have eked out an existence on the western and southern Cape coastal plains, forced by the low nutritional value of the plant life to depend increasingly on marine resources for survival. It is no coincidence that the world's first evidence that humans fed on marine life is to be found along this coastline, where shellfish middens date back over 100,000 years before present. It was precisely this intensive, protein-rich marine diet over thousands of years that is believed to have spurred the increase in brain size of the archaics, pushing them over a morphological and behavioral boundary into their modern forms. It is clear from the archaeological record that something significant began happening around 100,000 to 130,000 years ago. For the first time, we see sophisticated tools like bone harpoons making an appearance; the use of ocher as a cosmetic becomes fashionable, and the start of culturally modern thinking is evident. Only along the Cape coast does all of this come together in the archaeological record of Africa.

This conceptual leap around 100,000 years ago appears to have coincided with another shift in climatic conditions in the southern African subcontinent, which allowed these isolated populations to migrate northward into Africa once again. The seeds of contemporary human culture that may have sprouted during this phase of coastal isolation are suddenly found farther afield. Within the next 50,000 years, we see the world's first deliberate burial of the dead—a young boy is laid to rest in Border Cave on the frontier of KwaZulu-Natal and Swaziland some 80,000 years ago—and the first clumsy attempts at agriculture in the form of the deliberate burning of geophytes to stimulate seasonal growth at the southern Cape coast site of Klasies River Mouth some 75,000 years ago. From there it is but a skip in the grand scheme of time to the appearance of the first rock art engravings some 30,000 years ago, the use of ostrich eggshell beads, and the advent of the Late Stone Age culture.

To us, the population expansion of the moderns occurred at a seemingly remarkable pace, given that the dominant form of transport was still two feet. But 50,000 years is a long time, and it is conceivable that if generations advanced their territory by a few dozen miles in their lifetime, the old world would have been well traversed within this period. Sea levels at times were lower than they are today, which would have made the crossing into Europe possible along land bridges at both ends of the Mediterranean. It still remains something of a mystery how these early *Homo sapiens sapiens* settlers crossed the sea channels beyond Indonesia to arrive in Australia some 60,000 years ago.

But this is where it all started, I thought as I turned off the main road into the West Coast National Park. It is a relatively new nature reserve proclaimed by the South African government in 1985 to protect the fragile duneland ecosystem and internationally recognized wetland that make up the Langebaan Lagoon (Afrikaans for "long channel") complex. Among its extraordinary features is the profusion of bird life. Langebaan is one of the main southern destinations for migrating seabirds and waders from the Northern Hemisphere. The sight of hundreds of thousands of birds on the islands of Malgas and Schaapen, just beyond the mouth of the lagoon, is staggering.

The main road through the reserve winds through the scrubland onto a narrow spit of land that separates the crashing waves of the Atlantic on the left from the gentle lapping waters of the lagoon on the right. Unlike the "big five" game reserves that South Africa is noted for, there are few large mammals in this park. There are the occasional odd-looking ostriches and reintroduced antelope such as the eland and springbok. For the most part, however, West Coast National Park has the atmosphere of an empty landscape, though there would have been a greater number of wild animals during Eve's time: leopards, lions, and rhinos were common, as well as a large variety of antelope during the summer months. Hunting or scavenging for these would have brought a welcome dietary relief from the mussels, scallops, and beached seals that were probably the mainstay of her diet. As the road dipped toward Kraalbaai, on the lagoon side of the peninsula, I braked to allow a small tortoise to cross the road. In addition to sea life, it is certain that the early modern peoples who occupied this coast relied heavily on the abundant reptile population for food. I wondered whether a tortoise on its back on Eve's campfire was the prototype of the modern cooking pot.

I stopped at the parking lot at Kraalbaai, a few hundred yards from where the footprints were located. I'd been here several times before, mostly with Dave Roberts. Save for the few yachts moored in the bay to my left, and the hazy outline of the town of Langebaan several miles across the water, the vista before me would have been very similar to one confronting Eve 117,000 years ago, when her prints were made.

Part of my job is to bring hominid fossils to life, to interpret the ancient past in a way that is meaningful. This entails reconstructing the flesh and blood that once surrounded old bones. I try in this way to understand not just the mechanics of hominid evolution but the lifestyles and behaviors of our ancestors. The frustrating question that this process inevitably gives rise to, a question that exasperates all those who study this science, is this: At what point in our five-million-year journey from being an ancient African ape did we actually become truly human?

Given the fragmented nature of much of our fossil record, the further we go back in time, the harder it becomes. To hold the splintered remnants of a three-million-year-old tibia, for example, and to look for clues as to whether its original owner was more comfortable climbing trees or walking on the ground, requires more than just a thorough understanding of anatomy, muscle function, and a vivid imagination. It necessitates a conceptual shift in thinking in which the context of the find may be just as significant as the find itself.

But contextualizing fossil discoveries still leaves a significant limitation: the paleontological handicap of trying to understand the evolution of the hominid consciousness. The structure of a two-million-year-old pelvis may tell us whether an ancestral species of our genus had the capacity to run, but the bones alone give no indication of the range of emotions that he or she experienced in surviving the vicissitudes of the African landscape.

Eve's footprints elicited within me a response that differed from my usual reaction to fossils. There was always a level of excitement, of dealing with life as part of a bigger universe, with whatever remnant from the deep past I came across. But these footprints somehow were more real, more tenuous, more poignant. Finding fossil bones indicates that someone had to die. Eve's footprints were an image or snapshot of a *moment* of life.

When I first saw them embedded in the sandstone, I instinctively closed my eyes and conjured an image of Eve beyond the traces of fragmented fossilized bone or a row of statistically significant numeric images on a geneticist's spreadsheet: What would have occupied her thoughts as she made her way between the sharp crops of calcrete rock and the patches of succulents and thatching grass reaching the sand at the water's edge? Maybe, as I described earlier, she was returning from the beach after looking for dead seals and driftwood, which washed ashore during the rainstorm that was also ultimately responsible for preserving her footprints. She may have simply been going for a swim. Collecting firewood would probably have been one of her daily tasks; judging from the landscape and scarcity of tree cover, this would have occupied many hours of her day.

If we are to walk in the footsteps of Eve, and in so doing gain an understanding of ourselves, then we must step back in time. Not just to explore the mind of the footprint maker, that of her immediate ancestors or even theirs, but to journey back to the very origins of humanity, to the base of the family tree. For it is in the very roots of the human family tree that we will see the clues that make Eve so special. Going back in her ancestry to her ape-man origins is where we will find the physical and mental traits that would make you and I, *her* descendants, the most successful and dangerous beings in the history of the planet.

For decades East Africa has been accepted as the birthplace of humankind. This so-called East Side Story presumes that most critical events in hominid evolution in Africa occurred in the Great Rift Valley system. As important as East Africa is in the search for human origins, this view of human evolution is rapidly changing. Over the past decade or so new fossils, and new research on old fossils, is changing this perception. No longer do the sites in Ethiopia, Kenya, and Tanzania possess the only good evidence for the milestones in human evolution. Now fossils from southern African sites—such as Sterkfontein, Swartkrans, Gladysvale, and Drimolen—are fundamentally changing many of the ideas formulated for the East African prehistorical environment.

Most of the fossils of our earliest ancestors found in southern Africa come from just a few sites—nestled closely together only 15 miles outside of Johannesburg—in a range of dolomitic hills known as the Witwatersrand.

Anyone who has traveled to this part of the world will attest to South Africa's majestic landscapes. The central grassland plateaus are buttressed from the eastern subtropical lowlands and coastal plains by the mighty Drakensberg ("dragon") Mountains, and cut off from the harsh western coastline by the desert scrub of the Great Karoo. To the northwest lie the mighty Namib and Kalahari Deserts, where some of the last representatives of Late Stone Age civilization still eke out an existence from the arid lands as hunter-gatherers. These rugged mountains, deserts, and the scraggly coastal plains hold some important secrets of humanity's most ancient past, indeed this landscape may have been the stage on which our own genus *Homo* played its first act. The emergence of modern humans is in fact a late appearance in a rather long African play. In order to understand who Eve was, we must follow her tracks back into the distant past. Millions of years before she left her footprints on that Langebaan beach, her ancestors, the ape-men—or to give them their scientific label, the australopithecines—walked this African landscape.

In the Footsteps of Eve will try to cast light on the transition from not only archaic *Homo sapiens* to modern humans but also from ape-man to man-ape to human. Based on new fossil evidence, the strongest evidence of the speciation event that gave rise to our own genus may be found in South Africa's fossil beds. My own research indicates that the study of human origins has relied too heavily on the scrutiny of ancient hominid skulls and not enough attention has been paid to what happens from the neck down. Skulls are dramatic; staring into the hollow orbs of the eye sockets lends itself to philosophizing about our earliest ancestors. It is more difficult to get excited about a fossilized fragment of a leg bone, yet such a bone can speak volumes on how our ancestors moved about. My research, and that of my colleagues, on the postcranial anatomy of early hominids has thrown up a number of contradictory characteristics that suggest a far more complicated view of evolution.

I sometimes think of the proverbial biblical Eve as an analogy for the world's first scientist. She was motivated by an insatiable curiosity to taste the fruit from the tree of knowledge in defiance of the prevailing orthodoxy. Can one not describe modern scientific endeavor in the same way? Similarly, we'll explore the South Side Story of human evolution and begin our journey into southern Africa's past, walking together in Eve's footsteps to understand the mysteries of humankind.

Lee R. Berger
Johannesburg, March 2000

FIRST STEPS

F ossils can speak to us. As new technological and philosophical horizons are traversed, the boundaries between the living and the dead are being redefined. To an untrained eye, an inanimate piece of bone or stone is lifeless, an echo of primordial abstraction held in awe in the palm of the hand. But in fact, as messengers from a moment in the past, these pieces represent a reality that once existed.

Paleoanthropology as a science is defined as the study of ancient humans. Its scope and importance, however, extend far beyond that simple definition. Paleoanthropology is an example of one of the greatest privileges and one of the defining characteristics of being human: the ability to examine history, not only our own history but also the history of other animals and plants. While most other creatures rely almost wholly on instinct or behavior learned during the relatively short span of their own lifetimes, humans have the unique ability to examine the past, looking back generation upon generation.

At its most basic level, paleoanthropology is an extension of an interest in one's own ancestry, yet clearly at a vastly different magnitude. If we assume the average human generation is 20 years, then one's grandparents would be two generations removed; one's great grandparents, three generations removed. The Wright brothers first flew at Kitty Hawk in 1903, just under five generations ago. The American Civil War, which began in 1861, occurred about seven generations ago. The French Revolution, beginning in 1789, was only about 11 generations removed from today. The birth of Jesus was only 100 generations away in time, while the first of the great pyramids of Giza was built around 3200 B.C. That may seem an incredibly long time to most people, yet it is only about 260 generations removed from us today. If one stood a representative from each of these generations in a line, it would not reach the length of a football field. The first civilizations emerged only a paltry 325 generations ago; plants and animals were first domesticated some 550 or so generations back. Humans started systematically burying their dead about 45,000 years ago, at approximately the same time that the archeological record shows us that the first permanent artwork was created, which is only 2,250 generations away from us today.

Eve, stepping down over those Cape sand dunes 117,000 years ago, would have existed just over 5,000 generations ago. But she too was part of a long line of ancestors stretching back four million years in time, when the first of our forebears began an awkward ascendancy to walking on two legs. That is approximately 200,000 generations removed from where we are today, and 195,000 generations before Eve. As primitive as she may be by our standards, Eve is closer to where we stand on the road of human evolution than the first members of our genus *Homo*, who emerged approximately two million years ago. And even they were the result of another two million years—100,000 generations—of evolutionary experimentation in the ancient African landscape.

Our modern civilizations have been in existence for a pitifully short time. Yet in geological terms, this whole parade of humanity has evolved relatively quickly. To put this in perspective, if we had to condense the four-billion-year history of our planet Earth into a calendar year, our earliest ancestors would have emerged less than one minute to midnight on December 31.

Our species, anatomically modern humans or *Homo sapiens sapiens*, are able, if we choose so, to alter our actions in any situation based upon the accumulated knowledge and wisdom of those hundreds of thousands of generations that have preceded us. We can accomplish this because of a powerful brain and, perhaps more importantly, because of human language, which probably developed during Eve's era, about 5,000 generations ago. Our brain and our ability to communicate our thoughts through language allow us to analyze situations and pass on information at a level that no other animal can, not even our most closely related biological relatives, the chimpanzee.

To trace the footprints of human history back to Eve and beyond her to the dawn of our genus *Homo*, we have to follow a trail to Africa. Originally, we all emerged from a prehistorical African environment, shaped in particular by a series of dramatic climate changes that reached a climax between five and seven million years ago. It was during this time that our earliest forebears broke away from a shared lineage with the ancestors of the chimpanzee. This split marked the first evolutionary milestone on our long journey toward being human.

The steps from monkey to human have been well disguised by time. They've also been complicated by the number of competing candidates clinging to the branches of the family tree. In fact, the more fossils that come to light, the less our family tree appears as a magnificently tall redwood with well-defined branches thrusting toward the pinnacle of human achievement. Rather it resembles a scraggly thorn bush whose spiked and twisted interwoven limbs would be hazardous to unravel. Many a scientific reputation has been cut on the thorns of this human family bush.

Right at the base of this family thorn bush or tree, we share a common ancestor with the chimpanzee, which remains our closest zoological cousin, sharing 97 percent of our genetic makeup. This is a closer genetic relationship than exists between dogs and wolves. Humans and chimps—along with other apes and monkeys—are classified as anthropoids, the higher primates. Primates are generally tree-living creatures whose characteristics include clasping five-digit hands and feet, forward-looking eyes that have the ability to see in three dimensions, large brains, and complex social behavior. Along with lower primates—lemurs, tarsiers, and bush babies—our original mammalian ancestor was probably a shrewlike creature that scuttled around the Cretaceous forests near the end of the age of the dinosaurs, some 100 million years ago. Primates developed rapidly in the wake of the extinction of dinosaurs about 65 million years ago, within the forests of Africa and Asia in particular, giving rise to a number of different primate species.

The enduring mystery of human origins is why one member of this primate family evolved into the dominant species on Earth that humans are today. What led one lineage of African primates to become so much more successful than any other? We may have a rough idea of *how* it happened, but *why* should it have? Perhaps we can never really be sure. The study of human evolution is like tracking a moving target. Our prejudices and perceptions inevitably taint our understanding of the past. For this reason it is convenient to use the term hominid as a catchall phrase for living and extinct creatures capable of walking upright. This excludes apes but includes the ape-men, even those who may not have been part of the direct lineage that leads to modern humans.

The term hominids comes from the classification hierarchy developed by the 18th-century scientist Linnaeus. Living humans are defined as the species *Homo sapiens sapiens* (wise wise man), of the genus *Homo* (true humans), of the family *Hominidae* (hominids), of the superfamily *Hominoidea* (hominids and apes, excluding the lower primates), of the suborder *Anthropoidea* (monkeys, apes, and humans), of the order *Primata* (including lemurs, tarsiers, and bush babies), of the class *Mammalia* (animals who suckle their young), of the phylum *Chordata* (animals with a spinal cord) in the Animal Kingdom.

To understand the milieu from which hominids have emerged, we must understand ancient Africa. The key to Africa's conduciveness to human development and habitation is in its size and diversity. The continent is enormous. When one strips away the vast deserts, the icy mountain peaks, the mosquito-infested tropics and all the other places where humans cannot live, one is still left with a continent that has around one-third of the habitable landmass on the planet. Additionally, Africa has more of its surface straddling the equator than any other continent. The combination of size and equatorial exposure results in an enormous diversity of habitats, so that within Africa there exists almost every biozone on Earth.

These different environments are characterized by their own individual geographies, specific temperatures and humidity, altitudes, soils, and underlying geologies. These conditions determine the plant life within each ecosystem. Thus Africa contains an enormous range of plant material. The country of Ethiopia, for instance, has a wider variety of plant life than can be counted in the whole of Europe.

Diversity logically breeds change. It is reasonably straightforward, then, to see how the variety of plants at the bottom of the food chain in turn drives and defines the diversity of the animals that feed off them. With this vast vegetative splendor, it is not surprising that Africa has more different kinds of mammals than almost all the other continents combined. It is this diversity that spurred human evolution, providing the stimuli that drove the isolation of one line of apes, our earliest ancestors, to evolve over five million years into modern human beings. The stress of handling this diversity is what has driven evolution to the point that we dominate not just all land and marine environments on the planet, but have extended our range beyond the limitations of our biological sphere into space. This is quite an achievement for a little African ape.

Could human evolution have happened anywhere else? Bob Martin of the University of Zurich, and others, believe there may still be traces of early human evolution in Asia that, Martin says, have been largely unexplored. This is doubtful. Asia has been explored as much as Africa, but Asia does not seem to have the diversity that Africa does. Even though Asia's forests are home to apes like the orangutan, there is not the same degree of broken woodland in which the early hominids appear to have lived. Add to this the scientific cocktail of genetics and fossils, which reinforce the case for an ultimate African ancestry for humanity. The fossil record shows no evidence so far of hominids living outside of Africa more than two million years ago, while even claims prior to a million years are hotly contested or scientifically disputed. On the other hand, remains of hominids older than two million years have consistently been found in Africa. The genetic evidence corroborates this, with most molecular biologists convinced that humans arose in Africa.

Evolution doesn't take place at a constant, gradual pace over time. The nature of our planet doesn't allow that. Although on a day-by-day basis there appears to be little change in our environment, ours is a world in a continual state of dynamic flux. The surface of our planet consists of two mantles that float over the molten core. The apparent permanence of the landmasses and oceans around us is an illusion: They are continually moving. Over time they have changed beyond recognition as powerful forces deep within the core determine what happens above.

As long ago as 1620, the explorer Francis Bacon pointed out the similarities between the eastern coastline of South America and the western seaboard of Africa. However, it was only in the early 20th century that the geologist Alfred Wegener found the same rock formations on both the eastern "shoulder" of South

America and the western "armpit" of Africa, and concluded that they were part of the same ancient land formation. He went on to develop the theory of continental drift. It has been subsequently established that Africa was linked not only to South America some 100 million years ago but also to Australia, India, Madagascar, and the Antarctic, which together formed the giant supercontinent Gondwanaland.

The drift of the continents over the past two billion years is the backdrop to the evolution of all species. During the ages the continents have fused and split; the Earth's crust has been torn by the trauma of tectonic movements, causing the seas to advance and recede in a slow-motion geological drama. At times these movements have been both dramatic and imperceptible. They continue today, with the North American and European plates gradually moving apart at the speed at which a human fingernail grows. The junction of these two plates is known as the Atlantic Fault, and straddling them is modern-day Iceland, geologically one of the most unstable places on Earth. The last major earthquake in Iceland occurred in 1789, which saw part of the island drop by 19.5 inches in just ten days. Most of the changes, however, happen far more slowly. Among the youngest mountains in the world are the Himalaya, which have taken only tens of thousands of years to rise. These majestic peaks were the result of forces unleashed when the Indian subcontinental plate left its original moorings next to Africa and slid northward to collide with the Asian plate. Consequently, Mount Everest is still growing today.

Continental drift affects not only the shape of continents but also temperatures, sea currents, winds, and rainfall patterns. These changes inevitably have an impact on the most basic element in the food chain: plant life. Any change in the vegetation of a particular region in turn influences the lifestyle of all those who feed higher up on the chain. The changes wrought by continental drift occur haphazardly, making evolution appear to operate according to "pulses." This theory, first developed by Elizabeth Vrba, a paleontologist formerly with the Transvaal Museum and now with Yale University, holds that the extinction of life forms and the emergence of new ones are driven by sudden changes in the environment. Her research shows that sudden changes in global temperature at the end of the Miocene prompted such a pulse to occur.

The Miocene, a 15-million-year geological era that ended approximately 6 million years ago, was a time when Africa was carpeted by lush tropical rain forests. The thick woodlands offered ample food and safe shelter high in the trees, and the predominantly wet conditions that prevailed during the early Miocene ensured there was no shortage of water. These were conditions in which primates flourished. It was an environment that favored individuality. Twigs, berries, and leaves were there for an easy picking, so food gathering did not require complicated group behavior or social interaction. Then, during the middle Miocene, approximately five to seven million years ago, a new ice age began that had far-reaching effects on the African environment. The Antarctic ice sheets expanded, pushing cold water northward and causing a cooling of global temperatures. This was compounded by a continental shift that created the land bridge between the Americas, in turn changing the pattern of the ocean currents and planetary wind systems. Consequently, sea temperatures dropped between seven and ten degrees.

As agonizingly slow as this process may have been—taking several million years to occur—the net effects were dramatic. Land temperatures fell in the face of the icy winds that blew off the cold oceans. Sea levels plunged as more and more of Earth's water froze into ice. North Africa and Spain were connected as the Mediterranean began to dry up and become an inland sea. Africa escaped the icy conditions that gripped the Northern Hemisphere continents, but the net effect of this global cooling was the destruction of the extensive Miocene forests, which gave way to vast tracts of savanna grassland. Africa entered a lengthy period of aridification in which rainfall became more seasonal and erratic. The Pliocene, which began just over five million years ago, was the pulse that marked the culmination of this process, pushing some species into extinction and allowing others to emerge.

It is not difficult to see how forest-dependent creatures became increasingly vulnerable beyond the equatorial band of tropical rain forests. Dense woodland became restricted to river courses as water became scarcer and scarcer. Primates became increasingly isolated in this riverine bush, which proved to be convenient feeding pockets for new predators of the open grasslands such as lions and hyenas. There were also other dangers. Not far from Eve's footprints at Langebaan is a remarkable fossil-rich quarry in an old phosphorus mine, which is the site of a five-million-year-old riverbed. At one particular turn in that riverbed are the fossil remains of over 100 giraffids, precursors to the modern giraffe, which perished as a result of a flash flood prompted by the seasonal rains that were sweeping across Africa. Here the fossil record shows the disappearance of a number of Miocene species and the arrival of more modern forms of animal life as the subcontinent began to assume the ecology that is still with us today. Grazers such as the ancestral white rhino, *Ceratotherium praecox*, became more common, while specialized leaf-eaters like the African bear, *Agriotherium africanum*, disappeared. The widespread advance of the savanna encouraged a mass emergence of grazing animals such as various antelope species. Small pigs and peccaries, which lurked in the dense Miocene undergrowth, evolved into open-country species like the warthog.

On the fringes of these evergreen tropical forests, which today are restricted to central and West Africa, existed a number of populations of chimpanzee-like apes. As their traditional habitat altered inexorably, they were faced with considerable environmental stress, mostly because their feeding habits were disrupted by the climate change. Over time, the choice forced on these primates was either to cling to their arboreal lifestyles in the shrinking forests or adapt to the new challenges of a more open landscape with different kinds of food and danger. Perhaps there wasn't really a choice and geography was the determining factor. Those apes that lived in the heart of Africa's equatorial rain forests did not have to change because their environment remained stable; they just became more specialized to living in this environment. Today's chimps and gorillas are adaptations of large-bodied apes to tropical evergreen forests. Hominids, in contrast, are adaptations of large-bodied apes living everywhere except evergreen forests. In other words they became generalists. This is the fundamental difference between chimps (and their evolution over the past seven million years or so) and hominids, who developed from those apes that found themselves on the fringes of the forest. Faced by encroaching grassland, they had to adapt their lifestyles to a greater degree, a process that ultimately led to bipedalism, the ability to walk on two legs.

Although the exact trigger of bipedalism remains a matter of debate, it is generally accepted that the changing environment was an influential factor. These ancestral apes would have had to cross open spaces between the patches of trees that they still relied on for their food and shelter. At some point, walking upright became the adaptive locomotor strategy for moving effectively between suitable habitats. These open spaces of grassland would have been very dangerous for early hominids, who were easy prey for the lithe predators of the savanna. The main advantage of bipedalism was almost certainly that it freed the hands, allowing hominids to live in a three-dimensional environment instead of a two-dimensional one. In other words, they could interact with their environment far more expansively than they could previously, particularly while food gathering. All manner of new experiences thus became possible.

Kevin Hunt of the Peabody Museum at Harvard University is a scientist who has studied chimpanzee feeding habits extensively. He believes that 84 percent of bipedalism in chimps occurs when they are feeding—that is when they are on the ground reaching for fruit in the trees or moving between trees. Although at best this can be circumstantial evidence for an argument that early hominids adopted this posture for the same reasons, it is entirely plausible. The evolution of our upright gait was a process that probably took several million years to develop. The ancestral anatomy would not have allowed a rapid transition to two legs. Hunt believes it started with what he calls postural bipedalism, the scramble between clusters of trees, which eventually evolved into locomotory bipedalism, the consciously preferred way of moving around. Hunt says this could have developed between four million years, with the early australopithecines, and about two million years, when we see the appearance of early *Homo*.

It has been generally accepted that the climate changes that marked the end of the Miocene had something to do with bipedalism. There are, however, a wide number of interpretations as to exactly how the drier conditions affected early hominids, whose traditional arboreal habitat relentlessly retreated in the face of the advancing savanna. Could these early hominids have walked on two legs before the encroachment of the savanna, or was this the force that changed them? The notion that the savanna itself was responsible for bipedalism has generally fallen out of favor with evidence emerging that some form of upright walking existed before the grasslands dominated. Most scientists today accept that bipedalism arose in conditions of broken woodland where early hominids could integrate arboreal and terrestrial lifestyles.

English physiologist Peter Wheeler also believes that bipedalism arose from early hominids moving through the open grasslands between patches of forest, but argues that the most important function of bipedalism is one of controlling body temperature. Early hominids faced with increasingly high temperatures and a shortage of water had to literally find a way of keeping a cool head under the circumstances. By keeping the head high off the ground, it can be cooled by wind. Wheeler points out that humans are the only primates that sweat, a process that enables the body to lose heat and that indirectly helps maintain internal temperature control.

Owen Lovejoy, an American scientist who is widely regarded as the doyen of hominid locomotor studies, has a novel answer to the origin of bipedalism. While he, too, is firmly of the view that hominids had already begun walking upright before the encroachment of the savanna, he suggests that the origin of bipedalism is connected to sex. Not only does the bipedal posture show off the genitalia effectively, says Lovejoy, but hominids have the largest mammary glands and males have the largest penises relative to other primates. Herein lies the evolution of sexual attraction, and therefore reproductive behavior. Lovejoy argues that any development that increased the survival chances of the offspring would lead to new adaptations. Owen believes that walking upright would have enabled males to have free hands to gather high-protein foods such as eggs, grubs, reptiles, amphibians, nestlings, and worms. Not only could these foods be traded for sex, but they enabled the females to increase the time available to care for infants by reducing the amount of time they had to spend gathering, thus raising the chances of survival for the group as a whole. Lovejoy's thinking raises a number of questions about social bonding and the evolution of love. He may be criticized, however, for trying to use current perceptions of humans to explain a mode of hominid behavior that we can only imagine. Obviously, there are no fossils of hominid lifestyles, and although there was a social context to the interaction of our earliest ancestors, we must be careful about interpretations that are based almost solely on modern human behavior.

Bipedalism does not come without its disadvantages. There are obvious anatomical problems: the strain this form of locomotion puts on the back and the change in shape required of the pelvis, which in turn has made childbirth more dangerous for humans than for any other mammal. The added complication is that the bigger brains that humans have developed increase the physical difficulties of getting the baby through the birth canal. The result is that human, and indeed hominid, offspring have to spend much of their development outside the womb, which makes them exceptionally vulnerable and necessitates a longer period of infant dependency and parenting.

Whichever scenario one accepts for bipedalism first emerging, most scientists would agree that to a greater or lesser extent it was a result of climate change and the shifting of environmental circumstances. If you imagine that these ape-like creatures existed in the trees, it was literally an adapt-or-die situation as their habitats changed. The argument hinges on the assumption that two legs are better than four legs on a primate in an open environment and that a quadrapedal body is better suited to the trees. This is evident in the primitive characteristics associated with apes: very long arms, very long fingers, and toes that have a high degree of mobility and curvature to enable better grasping and use of tree cover. There is a dramatic shift in morphology when one examines a savanna lifestyle. Here hominids have adapted most dramatically from the waist down—in the pelvis, femur, the knee joint between the femur and the tibia, and of course the foot, which has been dramatically altered to sustain terrestrial bipedal locomotion.

Once some form of bipedalism and a new approach to food gathering had been established, the framework for subsequent evolutionary development was then set. There was an increase in brain size, probably through an improved diet, and the increasing sophistication of the social group, which some scientists argue was in itself a factor that stimulated early hominid brain development. The advantage created by a bigger brain, in this view, was not more efficient food-gathering skills but in being able to be more socially adept. The larger a community, the more the individuals within that group need socially manipulative skills. The disadvantage hominids face in having larger brains is the trade-off with the size and shape of the pelvis, the primary function of which is for walking upright. This limits the size of the birth canal, which in turn restricts the extent to which brain development can take place in an unborn baby. Too much brain development before birth would make the baby's head too big for the mother to give birth. The resolution to this mechanical problem is that much of infant brain development occurs after birth. This is only possible, however, if the social structure of a group allows for child dependency, that is for the mother—and possibly the father—to be relieved of certain group responsibilities in order to spend a lengthy time rearing their young. The effect of this is not only in the bonding that happens between males, females, and their young, but also in the way that it allows some form of education to take place. All of these factors involve sophisticated group interaction and cooperation that would have been a prerequisite for the development of language.

Most students of human evolution would agree with the broad brushstrokes of this picture of the early emergence of hominids. However, the devil is in the detail. Placing the fossil record within this framework and establishing the relationships between the fossils themselves are the subjects of continuous and usually heated debate, often with little consensus.

Presently, there are at least 17 recognized different hominid species that have existed in Africa between five million years ago and the present. Not all of them are part of the human lineage, and there is no definitive proof as to which of the early hominid species gave rise to the genus *Homo*. This proof may always be elusive, as each new discovery inevitably throws up as many questions as it does answers about the nature of early hominids and the relationships between them. Future discoveries have as much potential for creating gaps in the fossil record that we never knew existed as they do in providing a more complete record of early humanity.

Imagine a line of couples representing the 200,000 generations of hominids that have lived on the Earth since bipedalism began. From our comfort at the front of the line, it would be unnerving to walk back down the line to where the chimpanzee-like couple would be squatting right at the end. Having got over the surprise of how close to the front of the line we found our ocher-smeared Eve, we would have a very clear idea of how the process of evolution unfolded. The farther back we walked, the more we would notice brow ridges becoming pronounced, jaws protruding, noses flattening, physiques becoming more robust and stocky. Halfway down the queue, we would become uncertain about the relationship between animal and human in the figures we passed. Their bodies would be hairier, their smell would be stronger, their limb lengths would appear disproportionate; as we got closer to the back of the line, the difference in size between males and females would be extraordinary.

Walking down our imaginary queue, it would be relatively easy to find the defining moments of morphological change. But in reality we are faced with a different situation. Because these changes happened over hundreds of thousands of years and left tiny fragments of proof, the task of identifying the transition forms between species is a highly speculative exercise with a huge margin of error. Of the 400,000 couples in that line, our existing fossil record would randomly represent fewer than 5,000 individuals. And even then, not all of these individuals are part of the human lineage and would therefore not be entitled to stand in that line. In any other discipline, it would be precocious to use such a small sample to try and define a universe. The challenge of paleoanthropology is first to inspect the existing evidence and visualize the human family line, then to decide where—if at all— to place those individuals in the line up.

Although it is still woefully incomplete, the cast of significant characters in the hominid fossil record is substantially larger than it was a decade ago, due to the number of new discoveries. There is now general consensus that there were at least four broad categories of hominid that emerged after the split from the ancient chimpanzee lineage—although any new discovery has the potential to change this. The evolutionary pathway appears to have led first from a common ancestor with the chimpanzee to the genus *Ardipithecus* between four and five million years ago and then to the gracile australopithecines before the emergence of the robust australopithecines and our own genus *Homo,* approximately 2.5 million years ago.

The australopithecines ("southern apes") or ape-men, as they are commonly called, would have been huddled toward the back of our imaginary line of humanity. Besides the fact that they would have no conception of how to stand in a straight line, it would be very difficult to know which of them actually belonged in the queue and which did not. Compared to humans, they were small-brained, but their ability to walk upright and the number of anatomical features they share with us have placed them in the broad ambit of the human family. The variety among the eight or so known species is startling, a point that has made classification a contentious issue.

There is agreement, however, that the ape-men can be divided into two categories—the graciles and the robusts (who some argue should be recognized as a separate genus—*Paranthropus*, or "near man"). As the names suggest, the *gracile* skulls had finer facial features, while the *robusts* had enormous jaws and teeth. The australopithecines as a group emerged between four and five million years ago and, in the face of the changing African environment, underwent a number of evolutionary adaptations before disappearing into extinction approximately one million years ago. Exactly which of these ape-men gave rise to our genus *Homo* has not yet been proven conclusively, but the main contenders appear to be either *Australopithecus afarensis, Australopithecus africanus,* or possibly a species yet to be discovered.

The oldest known possible hominid is *Ardipithecus ramidus,* a 4.4-million-year-old fossil discovered in 1994 in the Afar region of Ethiopia by Tim White's University of California, Berkeley, research team. White believes that the creature *Ardipithecus*—derived from the Afar language for "ground ape," while "ramid" stands for "root"—is the first species on the human side following the split from the ancient chimpanzee lineage. "It's the link that's no longer missing," he told National Geographic Magazine in February 1997, but some of us are not so sure. Over nine field seasons, White's team found fragmentary remains of some 50 individuals—mostly teeth, bits of skull, and upper-limb bones. Preliminary assessments are that *ramidus,* who lived in dense woodland, would have weighed about 66 pounds and would have been just under 39 inches. Unfortunately, there have been few lower limb bones recovered so far, and until a formal description of the specimen type is published, the bipedal status of *ramidus* will be uncertain. White, who claims to have 100 bones from a single specimen, including an anklebone, is confident that *ramidus* could walk on two legs. But because *ramidus* represents a transition point in the adaptation from living in the trees to living on the ground, the way it walked would have been clumsy by today's standards: "Let's just say *ramidus* had a type of locomotion unlike anything living today," says White. "If you want to find something that walked like it did, you might try the bar in *Star Wars.*"

The first hominid that we can be sure walked on two legs is *Australopithecus anamensis—anam* is the Turkana name for "lake," hence "southern ape of the lake"—believed to have lived between 4.2 and 3.9 million years ago. Fossil remains, recovered by Meave Leakey's Kenya National Museum hominid team from two sites along the shores of Lake Turkana in northern Kenya, indicate that *anamensis* was about twice the size of *ramidus*, probably weighing about 121 pounds. More significantly, the *anamensis* tibia, or knee bone, is the earliest proof of bipedalism. The environment in which *anamensis* lived was riverine woodland, a more open landscape than the thickly forested areas that *ramidus* inhabited. Whether *anamensis* was a descendant of *ramidus* or an as-yet undiscovered mother species remains to be seen.

From about 3.5 million years ago, the fossil record starts becoming complicated by a profusion of species and a variety of interpretations, particularly with regard to where the thread of human ancestry passes. In East Africa there is the emergence of *Australopithecus afarensis* ("the southern ape of Afar"), whose most famous representative is Lucy, the skeleton discovered in Ethiopia by Don Johanson in 1976. The teeth recovered from the 3.3-million-year-old *afarensis* are far more humanlike than *anamensis*, and the postcranial bones—all bones other than the skull—show that *afarensis* was increasingly adept at walking upright. The difference in size between *afarensis* males and females—in scientific terms, sexual dimorphism—appears to have been enormous, so much so that questions have been raised as to whether remains found at one site are not in fact two different species. *Afarensis* males would have stood well over 3.5 feet tall, with a body weight of between 110 and 154 pounds. The females of the species were almost half their size, weighing only 44 to 77 pounds.

In North Africa another form of australopithecine has been located in this same time period. The relatively recent find of a mandible has been attributed to a new species, *Australopithecus barelgazeli*, believed to have lived between 3.3 and 3.1 million years ago. *Barelgazeli* appears to be a variation of *afarensis*, although only parts of a skull have been found, so nothing is known of the rest of its body. *Barelgazeli* was recently found by the French pale-ontologist Michel Brunet, suggesting that the early gracile aus-tralopithecines may have developed as regional variations.

At roughly the same time as *afarensis* makes an appearance in East Africa, *A. africanus* ("the southern ape of Africa") emerges in South Africa. Most *africanus* fossils found so far have been dated between 3 and 2.6 million years ago, and the younger specimens indicate that they may represent a species at the start of a transi-tion—either to *Homo* or to the robust australopithecines, or both. *Africanus'* skull appears more humanlike than that of *afarensis* but has a more primitive body. On the whole, *africanus* individuals were slightly smaller than *afarensis* and with less sexual dimor-phism. An *africanus* male, standing about 3.5 feet tall would have weighed between 99 and 132 pounds, while the female body weight ranged from 55 to 77 pounds. Both *afarensis* and *africanus* lived in broken woodland, but *afarensis'* larger body size may indicate that it was adapted to a terrestrial lifestyle while *africanus'* lighter bone structure supports the notion that it may still have been primarily a tree dweller. The most famous *africanus* specimen is undoubtedly the Taung skull, announced by Raymond Dart to a disbelieving world scientific community in 1925. The critical difference between *afarensis* and *africanus*—and one that will be discussed in greater detail—is that the former appears to have a more primitive skull and humanlike body, whereas *africanus* has a more advanced skull and a possibly more ape-like body.

There is strong evidence that between 3 million and 2.5 million years ago another evolutionary pulse occurred with large-scale changes in the global climate. The gradual warming in the wake of the temperature drop at the start of the Pliocene was radically reversed by an intensification of the ice age in the Northern Hemisphere. The ice that already covered much of North America and Europe crept southward, enveloping what would be present-day London and New York, while Siberia resembled a frigid desert. Because significant amounts of Earth's water were trapped in ice and could therefore not evaporate, rainfall patterns over the Southern Hemisphere continents were disrupted. Drier conditions in Africa led to a more extensive invasion of the savanna into the forest areas, putting more pressure on those species that relied on the woodlands for survival. By two million years ago African mammals, forest-dependent species, were—across east, north, and southern Africa—replaced by grassland-dependent species.

The response in the greater hominid family to these conditions is interesting. Two contradictory evolutionary survival strategies emerge: One involves specialization—those hominids that adapted themselves very specifically to the harsh new habitats, and the second involves a generalist response—morphologically opportunist hominids who did not become too dependent on any one habitat. The robust australopithecines were the specialists, adapting almost entirely to grassland food types, whereas early *Homo* was apparently a generalist, living and feeding in whatever habitat was most conducive to staying alive. In hindsight, human evolution obviously favored the generalists, who handled the stress of a changing environment more effectively and thereby crossed the cerebral Rubicon by developing bigger brains.

A remarkable window into this evolutionary adaptation is being studied at the world famous Sterkfontein site west of Johannesburg in South Africa. There, in a cave deposit known as Member 4, is evidence of a speciation event: fossils of one species exhibiting a shift in morphology that eventually turns them into another species. The speciation window at Sterkfontein has been dated to between 2.8 and 2.6 million years ago, fairly soon after the intensification of global cooling. The individual *africanus* specimens recovered there show an astounding anatomical variation, the implications of which are still being considered. It would appear that *africanus*—through the development of bigger teeth and more humanlike physiology—is being transformed either into the robust ape-man form or into true humans—or both! Although it is premature to speculate on what we will eventually make of these fossils, there is a very real possibility that a transition from ape-man to human occurred in southern Africa.

The first robust ape-men in our fossil record appear in East Africa in the form of *Australopithecus aethiopicus*. *Aethiopicus* fossils recovered from Kenya have been dated to 2.7 million years ago, possibly evolving from *afarensis* in response to the drier conditions and the deterioration of the tropical forest and broken woodlands of Pliocene Africa. In body mass and height, the robusts were slightly heavier than the graciles, but their facial appearance was completely different. All of the three known robust species had massive teeth and jaws, which they used to process the low-nutritional plant material of the savanna. The imposing sagittal crests the males developed to help root their huge jaw muscles to their skulls would have made them a fearsome sight.

From 2.5 to 2.3 million years ago, the fossil record in Africa is confusing, in part because of the number of new adaptations by hominids as a response to the 3-million-year-old evolutionary pulse. Not only are the gracile australopithecines giving way to a more robust version, but the first evidence of our own genus *Homo* occurs in East Africa, as well as an in-between hominid in the form of *Australopithecus ghari*. *Ghari*, which means "surprise" in the Afar dialect, is the latest ape-man form to have been discovered. Unearthed by Tim White in 1997 near the village of Bouri in the Middle Awash region northeast of Addis Ababa, *ghari* appears to be more advanced than *afarensis*. The description of the fragmentary skull and upper jaw still have to be published, but from preliminary accounts, *ghari* displays a mixture of gracile and robust characteristics; from the size of its teeth, it seems more closely related to *africanus* rather than *afarensis*. The intriguing aspect of *ghari*, which has been dated to 2.5 million years ago—and possibly the factor that surprised White—is that it has been found in association with stone tools, long thought to be an invention of *Homo*. These tools, made from rocks not found at Bouri, may have been carried from Gona, 59 miles away. Gona is the site of the world's earliest known stone tools, dating back 2.6 million years. Until the *ghari* findings are published, however, it would be idle speculation to suggest that this species was the first toolmaker or even, as some researchers are suggesting, the ape-man that gave rise to *Homo*.

Although the australopithecines probably made use of sticks, stones, and bones as ready-made weapons or as food-use implements, their ability to fashion stone tools is dubious. The manufacture of flakes and cores presupposes an intelligence in hominid groupings that had not been present before. It required abstract thought to visualize the design of a stone tool and a certain dexterity to transform that design into a functional object. Toolmaking must also have had a social imperative—the skills have to be passed on to others in the community, which would probably have strengthened bonds within the group.

Ghari may not have been the first toolmaker, though, as the earliest evidence of *Homo* also occurs at 2.5 million years ago, although not in the same locality. Don Johanson has found facial fragments of a *Homo*-type skull in neighboring Eritrea, in the Horn of Africa. A 2.3-million-year-old upper jaw found in Hadar, Ethiopia, in 1994 confirms the trend toward the emergence of true humans, distinguished by a larger brain, a bigger body, and a dental palate in which the teeth are smaller and set in broad, short rows.

Which *Homo* species was the first to develop? At this stage there is no definitive answer. Too little fossil material has been found of the two *Homo* discoveries mentioned above for them to be assigned reliably to any taxonomy. The first known *Homo* species are those of *rudolfensis*, discovered in Kenya near Lake Turkana when it still had its colonial name of Lake Rudolf, and *habilis*, the "handyman" first discovered by Louis Leakey at Olduvai Gorge in Kenya. Both of these species occupy dates of just over 2 million years before the present. Originally both were classified as *habilis*, but *rudolfensis'* bigger face and brain (about 800 cubic centimeters in capacity) warranted a separate species from the original *habilis*, which, with its smaller face and brain (650 cubic centimeters), is far closer to the gracile australopithecines. There remains some debate as to whether *habilis* should be classified as its own species at all, or whether it actually is an advanced australopithecine. These early *Homo* species, which may have been part of a number of short-lived evolutionary experiments, are collectively known as the habilines.

During the period in Africa just before two million years ago, the animals of the savanna, grazers, had replaced almost all the mixed woodland species, which were generally browsers. To survive the open plains, there was safety in numbers. Herds of grazers such as the ancestors of the wildebeest and hartebeest appear for the first time, while the variety of antelope increased, and zebras replaced the early browsing horses. The response by predators was also in numbers—lions hunting in packs became the most feared carnivores of the savanna, and the more individualistic saber-toothed cats disappear. It is fair speculation that hominid social structures became more sophisticated—for both food gathering and protection—and that this development was related in some way to the *Homo* characteristic of a bigger brain.

During the confusing hominid period between 2.5 million years and 2.3 million years ago, a number of protohumans occupied the African landscape. In East Africa there would have been populations of *A. ghari* and *A. aethiopicus* (which appears to be evolving into *A. boisei*, a type of robust australopithecine), *rudolfensis* and *habilis*, all competing for the same dwindling resources. In South Africa, a lack of fossil evidence from this period makes it less clear as to which hominids were present. Given the speciation window at Sterkfontein 2.8 million years ago, one can assume that *A. africanus* evolved into either a robust australopithecine, *Homo,* or both. What is clear is that by 1.9 to 1.8 million years ago, *A. robustus* has made an appearance at Kromdraai near Sterkfontein, while another hominid that has scientists puzzled also appears to have evolved from *africanus.* The scientific jury is still out as to whether Stw 53—its official title—is an advanced australopithecine or *habilis.*

How did these different groupings of protohumans react when they came into contact? Could they communicate through common sign language? Did they fight or did they try to avoid each other? Was there any form of interbreeding? The *Homo* populations probably had the upper hand when it came to commandeering resources. Their superior brainpower and ability to coordinate socially would have given them an advantage over other hominids in monopolizing the best campsites close to water and game. Competition between these groups may well have increasingly marginalized the robust australopithecines into more difficult ecological circumstances, leading ultimately, just prior to a million years ago, to their extinction.

The flurry of early *Homo* species played itself out by about 1.8 million years ago. In East Africa there is still evidence of *habilis* and *rudolphensis,* but they seem to disappear within the next 100,000 years or so; in what appears to be a startling evolutionary leap, we get the appearance of *ergaster* ("working man") at 1.75 million years ago. The best known example of *ergaster,* which really is an early form of African *erectus,* is the Turkana Boy discovered in 1984 by Richard Leakey and Alan Walker at Nariokatome on the banks of Lake Turkana. This strapping 12-year-old youth, who was 5 feet 3 inches tall at the time of his death, would have been well over 6 feet tall had he lived to be an adult. His athletic physique is proof that the preconception that humans have grown progressively taller over the millennia is wrong. In fact *erectus* would probably have been physically more powerful than any modern human being, with a body crafted to cope with the harsh conditions of the African savanna. Walker believes it unlikely that the Turkana Boy was capable of speech. Communication, he suggests, probably took place as a mixture of grunting and sign language.

H. erectus were probably the first hominids who really mastered hunting. For a start, their technology was superior to the crude Oldowan stone tools (named after their discovery at Olduvai Gorge) that had been unchanged for the previous million years. The Achuelian culture (named after St. Achuel in France, where such tools were first discovered) represented a major leap forward in early hominid thinking. One perspective is that the Acheulian culture was the first formal expression of rules being introduced into society. These rules had to do with the way stone tools—in particular bifaces—were made and with what they were made. Bifaces were handheld, pear-shaped handaxes, mostly between 3.9 and 7.8 inches long; they were broad, relative to their thickness, and were probably used as a butchering tool. *Erectus* also increased the number of different tools in use, from scrapers to cleavers; significantly, for the first time an element of style creeps into tool manufacture. The patterns cut into bifaces are probably the first use of symbolism, suggesting either the beginnings of a belief system or the first stirrings of an aesthetic sensibility.

With this intellectual and technological leap forward, *Erectus* soon dominated most of Africa, marking the transition from early hominids being victimized by their environment to mastering it. They were terrain specialists, choosing to live mostly in the riverine forests that cut through the savanna or in valleys and wetlands. Hippos, walking larders of fat, were plentiful and must have been an important Early Stone Age convenience food.

Part of *erectus's* dominance over nature was his ability to control fire. It is uncertain as to when early humans first learned how to make fire, but a fossilized hearth at Swartkrans near Sterkfontein in South Africa proves that *erectus* at least knew how to control it as long ago as 1.1 million years. Bob Brain of the Transvaal Museum made the discovery and believes that an *erectus* group probably capitalized on a bush fire caused by lightning when they took burning branches back to the cave and kept the fire alive. Whether the ability to control fire had anything to do with the disappearance of the robusts around the same time can only be a matter of speculation.

Erectus was also probably the first hominid to populate the world in a series of migrations from between 1.8 and 1.5 million years ago. The first nomads passed through the Middle East into Eurasia and China, while later populations entered Western Europe through the land bridge that still existed between North Africa and Spain. Their expansion parallels that of the African carnivores, and it is conceivable that these early migrants followed predators such as lions, either as a means of tracking game or to scavenge from their kills. This may even have led to better hunting techniques. *Erectus* eventually colonized most of habitable Europe and Asia, although large tracts of northern Europe would have been too cold to settle on. The one constant in the Diaspora is the stone tool kit: Acheulian bifaces only went out of fashion around 250,000 years ago, over a million years after they were first crafted in Africa. Their distribution shows that half the world's population shared and passed on the knowledge of making these large stone tools.

There remains some controversy as to when the earliest migration out of Africa started. Generally, this is thought to have taken place after 1.5 million years ago, with *erectus* reaching Java and China by about 1.2 million years ago. Researchers from the Berkeley Geochronology Center believe, however, that the fossil skull of an *erectus* child found in Java is 1.8 million years old, while in T'bilisi, Georgia, scientists claim to have an *erectus* lower jaw dated to 1.6 million years ago. Most early *erectus* sites in southern Europe date back to just over a million years ago. In Spain an adaptation of *erectus,* classified as *Homo antecessor,* has been dated to around 800,000 years ago. It is believed to have been a forerunner to the Neandertals who emerged in Europe roughly 300,000 years ago.

A frustrating development then manifests itself as far as the African study of human evolution is concerned. From approximately 1.5 million to 500,000 years ago—a period dubbed the "Million Year Gap"—the African fossil record for hominids becomes incredibly spartan. Yet signs of *erectus* are everywhere: The stone tools they used lie scattered widely over the African landscape, and there is plenty of archaeological evidence to show that the region was not barren of human occupation at the time. In particular, the diamond digging areas of the Northern Cape province have rich Acheulian sites, particularly at the confluence of two of South Africa's biggest rivers—the Vaal and the Orange— where a dynamic population existed between one million and 500,000 years ago. Some of the biggest collections of stone tools in the world are to be found in this arid landscape. The *erectus* camp-sites that have been excavated indicate the area would have been much wetter than it is now and game much more plentiful.

But there are precious few human fossils. The reason may lie in the culture of our science. The way in which the popular and scientific press covers human evolution is often misleading, and part of the blame has to fall on the shoulders of scientists. We have become obsessed with the pursuit of the oldest fossils. Story after story concentrates on the discovery of the most recent "earliest" and "oldest" fossil to lay claim to being our ancestor. This is a shifting target, because one is always going to find something a bit older and a bit more primitive than what has been dug up already. Just ten years ago, it was believed that the oldest ape-men were some 3.5 million years old. In a decade, we've revised our thinking and have made them a 1.5 million years older. It is all a bit like the proverbial tiger chasing its tail. The relentless pursuit of the oldest fossils has created gaps in the fossil record. If the science was to spend as much time trying to fill in those gaps as it does on chasing our earliest ancestor, we'd possibly have a more coherent understanding of where we have come from. This is partially what accounts for the million-year gap. The fossils are there, and no doubt will be found, because there are no known environmental reasons as to why they shouldn't be. By way of contrast, the fossil record in Europe and Asia is very rich in specimens from this time, because European and Asian scientists are looking for *their* earliest ancestors with great vigor.

From the few human remains that have been found in the million-year gap, it appears that in Africa a new morphology begins emerging from roughly 600,000 years ago, when *erectus* begins taking on the form of the archaic *Homo sapiens*. The skulls of Saldanha Man, found on the Cape West Coast, and Broken Hill Man, from Zambia—both dating back approximately 500,000 years—show an increase in brain size and a shift in the dentition and facial architecture that hint at the modern human form.

From half a million years ago, there are fewer gaps in the fossil and archaeological record as the transition unfolds between *erectus* and modern humans. There is also more evidence of cognitive planning and group coordination around food gathering. A good example of this is in mass hunting. The Magaliesberg mountain range, which traverses the northern part of Gauteng Province in South Africa, is a natural barrier to game movement across the plateau between the Vaal River basin and the Bushveld country to the north. Migrating herds of antelope have traditionally passed through a number of gaps in these hills to reach winter grazing grounds. Close by one such pass is Wonderboom Cave, on the outskirts of the modern city of Pretoria. It appears as if this cave was used as a seasonal hunting camp by the archaics who occupied it during the animal migrations. Game moving through the mountain pass was easier to kill than on the open grasslands. The dead animals were then carried back to the cave, where the meat was eaten and the skins and bones turned into clothes and weapons. Once the game migration had passed, the bands dispersed to return when conditions were once again favorable.

Unfortunately, the fossil record tells us little about how these early archaic *sapiens* organized their society, what belief systems and rituals they may have had, or what kind of relationship existed between male and female. South African archaeologist Bert Woodehouse muses: "Did they paint or dance, to what extent had their language developed as a means of communications? Answers to such questions are largely conjecture, but organizers of game drives must have needed some communication medium to facilitate planning. Men who became bedaubed in blood while cutting up animals may soon have drawn fanciful lines on their own bodies, while the careful stalking of prey may well have developed into dancing. Shouts of triumph may have gradually acquired the harmony necessary to justify the title of singing."

CHAPTER 2

AFRICA CALLS

〜

I t all looked so easy. I vividly remember staring at the chalk lines making their way across the lecture theater blackboard. The confidence and authority with which my Physical Anthropology 101 professor drew out the phylogony, or family tree, in front of the Georgia Southern University undergraduate class, left little doubt that we had cracked the mystery of human origins. She painstakingly wrote the complex Latin names of species on the board, each one with a date beside it, and beneath that, a list of the anatomical characteristics that made it unique. She then briskly drew the lines between the segmented species, showing their ancestral or descendant relationships.

The end result was a straightforward diagram representing an upside-down tree with only two main branches. Right at the top of the tree sat a triumphant Lucy—*A. afarensis*—representing the earliest ancestor of all remaining hominids. Below her, one chalk line led neatly to the southern African ape-man, *A. africanus*, and then on to the robust australopithecines. The line ended abruptly with *A. robustus*, signaling the extinction of this lineage a million years ago. This was a side branch leading to the main action—a slightly longer line that stretched ten inches and a million years from Lucy to *Homo habilis*. From there the chalk progressed to

Homo erectus and then forked: One fork went to the dead-end twig of the Neandertals, *Homo neandertalensis*; the other to the archaic *Homo sapiens*, which was just a short chalk mark away from the anatomically modern *Homo sapiens sapiens*. The professor explained clearly and patiently that bipedalism had emerged early in hominid evolutionary history and that as the line moved down the blackboard, brain size increased.

Looking at the segmented species linked by all those chalk lines, I felt a deep sense of satisfaction: We understood human evolution. The most difficult part would be learning all those complicated Latin names. The neat way in which all of the fossil hominid species and their anatomies fitted into this phylogeny was enormously impressive. Lucy embodied the dominant thinking of the time, that humanlike bipedalism—walking on two legs—had emerged very early in the evolutionary history of humans, while the heads remained ape-like until the emergence of *Homo*.

That human family tree, which I had studied as a student in the mid-eighties, was modeled largely on the work of two of the world's most respected paleoanthropologists: Tim White and Don Johanson. During the 1970s, they had turned the science on its head, rejecting the conventional wisdom that *A. africanus* gave rise to *Homo*. This notion had prevailed for almost 30 years, and began with the grudging acceptance by scientists in the late forties that *africanus*—then represented by the Taung skull and specimens from Makepansgat and Sterkfontein in South Africa—was the "missing link" between humans and an ancient African ape. Taung confirmed humankind's African ancestry and put South Africa on the map of international science. As more finds came out of that region confirming the existence of the ape-men, so *africanus* became the central pillar in the search for human origins. But this proverbial paleontological rags-to-riches story did not last. In the late 1970s, with a cold and clinical academic precision, *africanus* was thrown out of the human family tree.

In a paper innocuously titled "A Systematic Assessment of Early African Hominids," published in *Science* on January 26,1979, Johanson and White replaced *A. africanus* at the head of our family tree with *A. afarensis*, the species' whose best known representative is Lucy. Lucy is a 3.2-million-year-old skeleton that Johanson and his colleagues had excavated out of the Ethiopian sediments of Hadar some five years earlier. Not only was she 40 percent complete, a fantastic find in and of itself, but she seemed to capture the imagination of the world in a way no other find had before or since.

Johanson and White put together a convincing argument based on a thorough examination of cranial and dental material belonging to most of the early hominids. Their case was as follows: *Afarensis* and *africanus* shared enough characteristics to be lumped together in the same genus. However, *afarensis* was clearly the older specimen: It had been accurately dated to at least half a million years before the earliest *africanus* specimen and had more primitive dentition and a smaller-brained skull. Logic would dictate that in the absence of any other evidence, *afarensis* was the mother species that gave rise to *africanus*. But Johanson and White went event further. Lucy, they concluded confidently, was the true direct ancestor not only of all later hominids, but the direct ancestor of *Homo*. They believed that *africanus*, with its large back teeth, at best gave rise to the lineage of robust ape-man that was doomed to extinction.

A few scientists, including Richard Leakey, attempted to argue that the fossil record was too incomplete to reach such conclusions. Leakey, among others, also questioned whether all of the skeletons found at Hadar represented one species because of the variations in anatomical size. While Johanson and White argued that these were males and females of the same species, and that the males were larger, Leakey suggested that the larger specimens may have been an early form of *Homo*. Along with his parents, Louis and Mary, Leakey initially believed that the roots of our genus lay far deeper in time, once suggesting that *Homo* may have arisen around five million years ago. Although Leakey has since discarded that view, his skepticism about Lucy's status as the root of the human family tree may have been valid.

Nonetheless, Johanson's charismatic forcefulness and White's scientific fortitude eventually won the day in the literature and textbooks. By the time the dust had settled, the detractors of *afarensis* were intellectually outflanked and forced onto their academic back feet for the next 15 years.

At the time this scientific coup was being played out in 1979, my chief concern was passing eighth grade. As a boy living in rural Georgia, little did I know that one day I would find myself at the center of an intense debate about human ancestry—a proponent of *africanus* the underdog in the heavyweight fight against the ruling champion of human origins, *Australopithecus afarensis*.

I had already begun developing something of a passion for prehistory as a youngster, thanks to the privilege of an upbringing on a family plantation just outside of the town of Sylvania in rural South Georgia. I spent my youth scouring the sand hills and pine forests of the low country, searching the ploughed fields for Native American potsherds and arrowheads. The sense of child wonder at coming across these artifacts would lead my imagination into a lost world. I'd sit for hours among the thick stands of pine and scrub oak, conjuring up images of the men and women who had camped, hunted, and fished in the places that I'd walked. I knew even then that I wanted to be a scientist studying prehistory. As I grew older my interest shifted. My chosen path in science was to be a geologist, and I intended to study the upper Cretaceous carnivorous dinosaurs, huge wondrous beasts like Tyrannosaurus, which lumbered around the undergrowth 70 million years ago. But by the time I enrolled at the university, I decided that people interested me more. I wanted to learn about the origin of the hominids, the earliest members of our lineage.

Since those days at Georgia Southern, the world of paleontology has again been turned upside down. A number of new hominid discoveries have been made, existing fossils reappraised in light of new techniques, and there has been a tremendous advance in the technology of genetic science, all of which have forced us to concede one main point: Human evolution is a lot more complicated than had been previously thought. The confidence with which my anthropology professor wielded that chalk across the blackboard was entirely misplaced. It wasn't her fault. Johanson and White and other scientists made it look so easy. That's not to say they're to blame for giving us wrong information about our evolutionary roots. They put forward a logical, well-reasoned argument, using all the data at their disposal, along with several well-intentioned educated guesses that *their* hominid, Lucy, was the mother species of all that followed. Since then, at least two species older than Lucy have been discovered, and my decade of researching human origins in Africa has convinced me that *afarensis* may not be a direct player in the human lineage. This book will present those fundamental contradictions and make a case that if we are to be so bold as to identify a mother species for *Homo*, *Australopithecus africanus* is a far more suitable candidate. Lucy may well have to relinquish her position as the mother of us all.

Research on the postcranial anatomy of early hominids—my particular area of study—has thrown up a number of anatomical anomalies in early hominids that suggest a far more complicated view of evolution than Johanson and White's original phylogeny. The study of human origins has relied too heavily on the scrutiny of ancient hominid skulls. Skulls are captivating specimens; the postcranial elements less so. The biggest problem, however, is in reconstructing the right relationships between the skull and the rest of the skeleton.

In the years ahead, there may well be another discovery that will invalidate the argument that *africanus* is the mother of our genus *Homo*. But that is the nature of the science; it's not an end-game in which we will all wake up one day and understand all there is to know about human evolution. Somewhere, somehow, another fossil will pop up and force prevailing theories to be reexamined yet again. So, with this in mind, I take on White and Johanson with humility. As a former disciple, I'm grateful for what they—and my other mentors, Phillip Tobias and Richard Leakey—have taught me.

In fact, I owe much of my good fortune to Johanson. During a weekend we spent together in Savannah, Georgia, while he was on one of his lecture tours, he gave me some controversial but useful career counseling. At the time, I was a postgraduate student and absolutely in awe of him. He was, in a manner of speaking, a rock star with a natural ability to demystify the com-plexities of the science and make it accessible to the broader pub-lic. His fastidious approach has earned him the respect of much of the scientific community and the thousands of readers who sub-sequently bought his book, *Lucy*. Johanson had me enthralled that weekend as I listened to him describe the excitement of the field team the evening after they had found the first pieces of Lucy. They knew that they had come across something special, but were not sure yet how special. They discussed the possibilities of the discovery and raised their beers in mutual congratulation as a tape of the Beatles hit song "Lucy in the Sky with Diamonds" played over and over again into the starry African night. That is, of course, how Lucy got her name.

I listened to him talk and was hooked. I wanted more. I told him it was my lifelong dream to go to Africa to search for hominid fossils. He chuckled, probably having heard the same thing from dozens of eager students, but then he gave me a singularly important piece of advice. There was little room in East Africa for newcomers because of the intense rivalry between scientists in the region and the governments' tight control over the issue of digging permits. "On the other hand," he said, "there's South Africa."

I was taken aback at the suggestion. South Africa at the time was a pariah state, teetering on the edge of a civil war as the minority government sought ever more desperate measures to maintain its power. Although in 1988 it was clear that changes were imminent, they were not yet a reality. Sanctions were still in place and most countries on the African continent would refuse entry to anyone with a South Africa visa in their passport. The debate about humanity in South Africa was not about the origin of the species but about the abrasive dynamics of contemporary human behavior inasmuch as it related to class and race, privilege and oppression.

Johanson's suggestion to work in South Africa presented not only an obvious moral dilemma but also a practical one. There were only two major international universities in South Africa, the University of the Witwatersrand in Johannesburg, also known as Wits, and the University of Cape Town. Of the two, Wits was the dominant institution for studying human origins, with research led by the great Phillip Tobias. Were I to be accepted at a progressive institution such as Wits, working under Tobias—himself a public opponent of apartheid—might still be damning in the eyes of my colleagues. It would also fly in the face of the United Nations academic boycott of South Africa and potentially damage my professional aspirations for working with people like Richard Leakey, one of the most ardent activists against white supremacy in Africa. I was very much aware that if I went to South Africa, I might never be allowed on Kenyan soil. Still, Johanson's suggestion had merit. He told me about a wealth of fossil hominids in Tobias's safe that the world simply did not know about. The fossils were yet to be described, and of course there was the potential to find other important fossils. After careful research and a good deal of moral soul-searching, I decided to take his advice and apply for a position at the University of the Witwatersrand in Johannesburg. Two dominant factors outweighed the considerable political risk in following this course of action: first, the opportunity to work with one of the greatest living paleoanthropologists, Phillip Tobias; and second, the chance to study original hominid fossils. Tobias had the reputation for taking only one graduate student at a time, and it was my understanding that there was presently a niche waiting to be filled in his program.

In the meantime, Johanson invited me to join a seasonal field trip planned for Olduvai Gorge. I was thrilled. *Olduvai Gorge*—just the sound of it had an exotic ring. The fossil fields of Olduvai had yielded some of the most famous hominid remains, and who knew what secrets still lay buried beneath those sunbaked sediments? Geographically, Olduvai Gorge is in northeastern Tanzania, but for me it lay at the center of the paleontological universe. Unfortunately, the promise of early fortune that I had fantasized finding at Olduvai would never materialize. No matter where Olduvai may have been on a physical or cosmic map, it was located firmly within the Leakeys' orbit of influence. And this simple fact stopped Johanson's planned expedition there.

Richard Leakey. Probably the most powerful person in African paleoanthropology today, he has a fearsome reputation for being stubborn and competitive—and also for being an exceptionally good fossil finder and mobilizer of resources in the search for human origins. He and Johanson had fallen out during the late 1970s in the wake of Lucy's discovery, and what might have been a mere professional disagreement materialized into personal animosity. This is understandable. At the best of times, paleoanthropology, because it is such an interpretative science, leads to personal rivalries. Entire careers can hang on a discovery, reputations can be made or destroyed in the subsequent critical analyses, and funding more often than not follows personalities rather than institutions. And, some would say, the study of human origins has more emotional resonance for its participants than, say, the science of microbiology. That's because paleoanthropologists are essentially studying humans. A debate about who we are and where we come from is likely to get heated. Paleoanthropology is also a science that attracts strong personalities, and both Richard Leakey and Don Johanson are forceful characters.

Ostensibly, their rivalry was rooted in the disagreement over the interpretation of Lucy. Richard Leakey was critical of Johanson's claims that *afarensis* was the original hominid ancestor. He also believed that Johanson was deliberately ignoring the possibility that Hadar, the site where Lucy was found, also contained another species, one that could have been an older form of *Homo*. Were this the case, then Leakey's views of a far more ancient emergence of our genus would have been validated. It was Richard's mother, Mary Leakey, however, who developed the most intense dislike of Johanson. She was upset that Johanson had used some of her findings from her Laetoli excavation in Tanzania to argue the case for *afarensis*. This was after Johanson had declared that Lucy's skeleton was the holotype for *afarensis*—that is the first or best specimen that defines a species—and then claimed that a mandible from Laetoli was the paratype—the next best specimen that defines a species. Mary shared her son Richard's reservations about the significance of *afarensis* and was outraged that Johanson was using her published work to back up arguments with which she did not agree.

"How can you even associate yourself with that man," she once scornfully chided White after he accepted the invitation from Johanson to describe the Hadar fossils. White, who had played a major role in analyzing aspects of Laetoli, soon found himself drummed out of the Leakey camp and unwelcome at the Leakey excavations. Mary Leakey went on to criticize White as naïve and dismissed Johanson's book on Lucy as a pander to populism, going so far as to accuse him of fabricating a number of conversations, which he had in fact quoted verbatim.

But the real reason behind their rivalry was probably that Johanson and the Leakeys had staked their reputations on finding the oldest human ancestor, and neither wanted to be pipped at the post. Richard Leakey also considered East Africa his personal stomping ground and saw no reason to share it with any outsider. He admitted as much when he confessed to undermining the efforts of rival paleontologists, one such instance being when he feared that a team working in Ethiopia's Omo Valley might encroach on his own fieldwork north of Lake Turkana. Leakey secretly told the Ethiopian government that the quality of their work was not up to standard, which resulted in stymieing their efforts. Leakey later candidly confessed: "I wanted it myself. And I was then an absolutely miserable fellow. I'm sure I was capable of being most unpleasant and uncharitable; I wouldn't have earned my reputation if there weren't some truth to these stories."

Leakey had to concede, though, that Lucy was a significant find that justified Johanson's international profile and growing influence in African paleoanthropology. There is an African proverb that aptly sums up their stubborn competitiveness: "You cannot have two bulls in the same kraal [cattle enclosure]." Although both Richard Leakey and Donald Johanson accused the media—with some justification—of blowing their professional rivalry out of proportion, they subsequently quietly carved out their own areas of influence. Kenya and Tanzania remained part of the Leakey empire, while Johanson and White made Ethiopia their "kraal."

So it was with some trepidation that I approached Richard Leakey to ask his opinion of Johanson's suggestion that I should go to South Africa. I had eventually succeeded in getting to East Africa after Johanson's Olduvai offer had fallen through. My application to Harvard University to join the 1988 field season at their Koobi Fora field school at Lake Turkana in Kenya had been accepted and my opportunity to meet Leakey came several weeks after I'd been at the camp. By this time, he was turning his attention more to conservation management than paleoanthropology, although he remained a vital force in the profession. One day he flew into the camp with visitors. I was slightly nervous about the prospect of meeting him, given his reputation, which was probably best summed up by Smith Hempson, the U.S. ambassador who served in Kenya between 1989 and 1993. Leakey, he said, was "ruthless, short-tempered, arrogant and self-promoting. He is also articulate, intelligent, unsentimental, industrious, and attractive to women, a workaholic reluctant to delegate authority."

Although I cautiously approached Leakey about his opinion of South Africa, he took considerable time to discuss the issues with me, and he was frank about the situation. Pointing out that although the winds of change were blowing in South Africa, it was still a pariah state and just about anyone associated with it would find themselves isolated.

"If you go, you'll probably never be able to enter Kenya again."

I looked out at the lake. Until that moment, I had not fully grasped what the professional consequences could be if I went to South Africa. What Leakey was really saying was that not only would I be banned from Kenya, but also from Tanzania and Ethiopia as well as a dozen other African countries. If South Africa did not work out, my planned career as a paleoanthropologist in Africa would be dead.

"Would you go in my position?" I asked, not being able to look directly at him.

"If it's original fossils you want to work with, the answer is yes."

Richard Leakey convinced me. I would go to South Africa if I was accepted into the Wits University program. I returned to the United States from the Koobi Fora field season excited and unsettled. Two of the most influential personalities in the pale-oanthropological profession had privately given me the same advice: Go to South Africa if you are serious about hominids. The fact that Johanson and Leakey were barely on speaking terms with each other—and yet concurred individually on the course of action I should take—was deeply reaffirming. I was prepared to risk the political consequences of going to South Africa on the reasoning that the situation there was changing, that it was a matter of time before apartheid would be abolished, and that there was virgin hominid territory to be had.

Besides convincing me that my future lay in South Africa, the Koobi Fora experience had inflamed my passion for the science: It was there I found my first hominid. I had been in a state of high excitement since I'd left the United States en route to Kenya. By the time I arrived at Uhuru airport in Nairobi, I felt as if I could burst. I had spent the flights and layovers rereading Richard Leakey's *Origins,* as well as Johanson's *Lucy,* for the umpteenth time. I would finally get a chance to walk on the very soils from which our ancestors had risen and in which these giants of the science had searched. Despite my healthy skepticism about religion, there was a sense that this was a spiritual journey for me: I was on a pilgrimage to the birthplace of humankind.

My appetite for the African landscape was whetted by a week of sightseeing in game parks. I marveled at the rolling savanna grasslands, where herds of antelope, wildebeest, and zebra mingled. The series of classroom lectures that followed told us there is comfort in numbers—the savanna favors the crowd more than it does the loner. The hominids contemplating life in the broken woodlands and grasslands during the great aridification of Africa from about five million years ago would have been forced by their environment to cooperate in order to survive. While an arboreal nuts-and-berries existence lent itself to an individualistic lifestyle for the primates of these forests, food gathering on the plains was a different story. To eat well was to learn how to communicate, to share, to work together, and to compete. This was the environment within which humanity played out its bipedal apprenticeship.

After the lectures, we prepared to depart for Lake Turkana, one of the world's most famous fossil fields, deep in the Kenyan interior on the frontier with Somalia and Ethiopia. After a two-day, 400-mile bone-jolting ride in the back of a reconditioned army lorry, we crossed a vast field of lava boulders and came to a halt on a rise overlooking a spectacularly dramatic lake. The emerald green waters of vast Lake Turkana disappeared into the distance. Ringed by ancient volcanoes in a seemingly desolate landscape redolent with a sense of history and destiny, Lake Turkana seemed to stretch out forever and ever. Although the mighty Omo River deposits all the water from the Ethiopian Highlands into this huge catchment area, there is no outlet to the 200-mile-long, 30-mile-wide lake. And yet its level is steadily declining. That's because all the water is lost through an enormous rate of evaporation. I took a deep breath as I absorbed the scenery and considered the anomaly. We still had a dusty six-hour drive along the east side of the lake before we could reach, by sunset, the Koobi Fora camp. Sited at the base of a peninsula just 300 yards from the edge of the expansive lake, the camp was more humble than one might expect. It consisted of a main thatch-roofed building with open-air sides and a series of smaller sleeping huts decked out with hammocks. On the nearby grassy plain, a number of antelope and zebra could be seen coming down to the water to drink. I felt I had arrived at the threshold of my dreams.

I spent my first night at Koobi Fora restlessly listening to the sounds of Africa. At 4:30 a.m. I could no longer lie still. I quietly got up and picked my way through the unfamiliar darkness to the main hut where a single solar-powered neon light shone.

There, sipping a cup of tea, was John Kimengitch, a heavyset Kenyan who was a famed member of Richard Leakey's "Hominid Gang." A genuine fossil-finder. He was amused at my early rising and pointed to where the kettle had just boiled. I poured myself a cup of tea and sat down at the table for an introductory chat. Then Kimengitch casually invited me to join him at a nearby fossil-bearing area. My heart leaped. There were no formal activities planned for that day at the field school, so I eagerly clambered into his battered Land Rover. As we eased our way slowly across the rocky landscape, the first light crept quietly across the jagged backdrop of hills.

Fossil-hunting satisfies the romantic instinct in that it usually takes place in some of the most beautiful and remote landscapes on Earth. But the impressive scenery soon blurs on the edge of one's peripheral vision because while looking for fossils, your time is spent staring intensely at the ground. It also gets hot. The reason why fossils are usually found in these harsh, arid regions is that year after year natural erosion causes the rocks to crumble, exposing previously covered ancient sediments.

I had searched for fossils in conditions similar to the Lake Turkana sediments in the badlands of South Dakota, but I was astonished by the volume of ancient bone that littered the landscape. There were eroded areas and little hillocks of gray sand where fragments of fossil bone literally covered the land. Although I had learned the principles of African mammalian bone identification, and had taught myself from books and casts what a hominid fossil should look like and how to differentiate it from other bones, I was unprepared to search through such a density of fossil bone on the surface sites.

Fortunately, Kimengitch was a patient teacher and demonstrated how to scan the bones on the surface, not necessarily looking at every fossil but for telltale clues that make one bone more interesting than another. He pointed out that hominids have straighter long-bone shafts and that the cortex of the bone is more thickened than in other animals. Look out for the thicker tooth enamel, he said; hominid teeth are more bulbous than the flattened, straight teeth of herbivores. My confidence levels rose as he gave me other identification clues, and pretty soon I was seeing the right shape patterns. Carnivore remains began to separate themselves from antelope remains, and both from the rare monkey fossil I would pick up. Kimengitch also taught me how to think about my identifications.

"At first," he would say in his soft Kenyan accent, "don't try to jump to a conclusion and figure out what the fossil that you have found is, but eliminate what it isn't. Try and remove the more common possibilities like antelope and carnivore before reaching a conclusion."

Occasionally I found a piece of bone that looked more promising than most and would call Kimengitch over for an opinion. The result was always the same: a gentle no and a detailed explanation of why not. By 11:00 that morning it was too hot to continue. As the sun beat down on my neck, I felt both exhausted and elated. As we made our way back to the vehicle, I felt a twinge of disappointment that a complete skeleton of a hominid had not popped up before my eyes. How unrealistic, I chided myself. To this day, though, every time I go into the field and don't find a fossil, that same feeling comes over me. Nonetheless, I was feeling pleasantly attuned to the process necessary for discovering these elusive fragments of deep prehistory. I was engrossed in reflecting on the morning's activity when a small fragment of bone caught my eye. I kneeled down beside it. It was a piece of a long bone, about four inches in length, with a straight shaft and a thickened cortex. I looked up for Kimengitch, who was about 50 yards away nearing the Land Rover, and wondered whether I should call him. This bone seemed to fit the shape of a central piece of a hominid femur shaft, the thighbone, but I had been wrong so many times already that day that I hesitated.

"John," I called, "can you come and take a look at this?"

Kimengitch turned on his heels and walked back to where I crouched over the small bone fragment, which I handed to him.

"That looks good," he said nonchalantly. "Congratulations on your first hominid."

I was dumbstruck as he handed the fragment back to me and shook my hand.

John collected stones for a small cairn to mark the site and pricked a pinhole in the aerial photograph he carried to mark the location. This was so we could return later to collect the specimen with Craig Feibel, the geologist with the field school, and the only person in our party authorized by Leakey to collect hominids. As John piled the rocks, I stepped back to take a photograph of him marking *my* fossil find. When I stopped to take the picture, I glanced down and saw another interesting fragment. I kneeled to pick up the bone—and couldn't believe it! This fragment was about three inches of the shaft of a hominid humerus. I called Kimengitch over again, and he just shook his head and grinned when I showed him the piece.

"Yes, that's another one."

After searching the area for another hour, we gave up on finding any more pieces. By this time, the sun had made the conditions almost unbearable as the heat rose off the rocks around us. But I barely noticed the discomfort. Those two little bone fragments had decided my future.

Back at camp that evening, I sat on the edge of the lake with my field journal and contemplated my luck. I looked out over the water and thought about some of the men and women who had molded our perceptions of our own species. They were part of an exclusive club that I wanted to join. After writing in detail about "my" hominid find, I recorded in my field journal what only a naïve 23-year-old anxious to jump into the science could write:

Watch out Richard Leakey.

I returned to Georgia in 1989, after the field season,

convinced that I was ready for South Africa, but the nagging question was whether South Africa was ready for me. There was no response yet from the University of the Witwatersrand. Perhaps there were no postgraduate vacancies left. I couldn't wait any longer and decided to phone the university. Because of the time difference, I had to wait until almost midnight before I put a call through to Johannesburg, where it was still late afternoon. I was in luck. The secretary in the anatomy department informed me that Professor Tobias was in and would take my call. Clutching the receiver tightly, I held my breath until I heard Tobias's voice. I then proceeded rather breathlessly to introduce myself, and asked whether any decision had been made on my graduate application. There was a momentary transatlantic pause on the line before Tobias responded crisply:

"Dear Boy, we would be very happy to have you here. When can you come?"

Just over a month later, I was on my way to South Africa. Predictably, the country was in desperate political shape. The white government was still clinging to power in the face of widespread domestic and international pressure. Nelson Mandela was still in jail. All the major anti-apartheid organizations were banned, and there was a state of emergency in force restricting civil liberties. A low-intensity civil war was underway between the African National Congress (ANC) guerrillas and the National Party (NP)–controlled security forces, a conflict neither side looked capable of winning decisively. On the surface, Johannesburg was a bustling city, far more developed than Nairobi and surprisingly First World in its infrastructure. Although the end of apartheid was apparently at hand, the vestiges of racialism were still everywhere. Signs stating, Slegs Blanke ("Whites Only") were commonplace in shops and restaurants. Even the English-style double-decker buses were divided along racial lines, with whites on the bottom and "other" groups on top. Apartheid structures were bizarre.The Japanese, I was told by colleagues, were given "honorary white" status on purely economic grounds, reportedly to give them access to the numerous country clubs.

The consequence of apartheid had philosophical as well as political implications. The NP had developed and refined their policy of racial segregation over a 40-year period to benefit its white Afrikaner support base. To justify this social engineering and the denial of the franchise to anyone who was not white, the ruling party leaned heavily on its own customized creationist perspective. Its apologists found oblique biblical references to argue that whites were genetically superior to all other races, and anything that threatened this worldview was treated as a heresy. The disciplines of human biology and paleoanthropology, which had dispelled the myths of racial hierarchies, were frowned upon. Evolution was not a subject taught at schools.

The black Africans I met on my arrival were surprisingly reconciliatory, a point that would be proven by the relatively non-violent transition to power that would occur a few years later. I moved into a mostly black apartment block in downtown Johannesburg that provided both safe and cheap accommodation, and was warmly embraced by my neighbors. The contrasts that surrounded daily life in South Africa extended to my home for the next ten years, to Wits University, and more particularly to the world-famous medical school. Wits Medical School lies in the concrete embrace of the enormous Johannesburg General Hospital, a blocklike, postmodernist architectural monstrosity that grimly dominates Parktown Ridge, which overlooks the city's verdant northern suburbs. Despite the school's underground bunker aesthetics, it housed a surprisingly friendly fraternity of staff and students, who prided themselves on their custodianship of the most sophisticated medical academy in Africa.

Wits was also the castle of the king of South African paleontology, the man who helped shape the way I think and to whom I owe so much of my science. Professor Emeritus Phillip Valentine Tobias, FRS, is a sprightly septuagenarian who has played a formidable role in the development of the science of paleoanthropology. His thinking has influenced almost half a century of science since he graduated from the University of the Witwatersrand Medical School in 1946 with a Bachelor of Science degree and, in 1951, with a Ph.D.—and his personal achievements have been astounding. He holds no less than 11 honorary degrees, and he has had at least 21 international medals, prizes, and awards conferred upon him, including Switzerland's prestigious International Balzan Foundation Prize. Tobias has been invited to lecture in more than 50 countries; probably most impressive is that he has authored close to 770 articles in every major scientific journal. He was Dart's pupil and successor, and ever since he was invited by Mary and Louis Leakey to describe first *Zinjanthropus boisei* and later *Homo habilis*, Tobias has enjoyed his status as a world authority on the early human record.

My first encounter with him was memorable. He took me to the university's hominid fossil vault and produced a large skeleton key from his pocket. As he inserted it into the heavy gray metal door, I held my breath. This was the Aladdin's cave of paleoanthropology, the reason that I had traveled halfway across the planet. Behind this door lay not only the largest single collection of undescribed early hominid fossils in the world, but my future. When the door was fully open, I looked into the safe. It was a room about ten feet long and six feet wide, with row upon row of small boxes containing fragments of bone or teeth.

"What do you think?" Tobias asked, looking at me with a knowing glint in his eye.

I was speechless. It felt like I was being let in on one of history's best kept secrets.

Most of the fossils, the bulk of which came from Sterkfontein, were unknown to the broader scientific community. For a variety of reasons, including the academic boycott, Tobias had not published much about this collection, which in turn meant that nearly 500 specimens were undescribed. At that time there were maybe 4,000 cataloged hominid specimens in *all* of the collections of both East and southern Africa. The skulls, jaws, partial skeleton, and other bones that were neatly stacked on those shelves would increase the known data sample by almost 15 percent! Johanson had not done the assemblage justice when he told me back in Georgia there were "a lot" of new specimens.

"Well, my boy, somewhere in there lies your future." And with a fatherly tone to his voice, Tobias gestured into the recesses of the safe and added, "There's enough in there to change the way we view human evolution."

Those were prophetic words. Over the next five years, those specimens proved to be the ammunition that would enable me to challenge the paleoanthropological orthodoxy of Johanson and White. My gratitude to Tobias was immense. But I soon began to realize that he was a complicated man. His quaint Old World gentlemanliness, charm, and twinkling eyes hid a ruthless determination that characterized his quest for scientific truth, academic recognition, and personal glory. I was puzzled when I first heard the whispered warnings in the corridors of the medical school to be careful of him. "Beware Tobias, he eats his young," a colleague remarked theatrically over a cup of coffee in the lab. I resolved not to take sides in any departmental rivalry and privately pledged my loyalties to the science as a whole. Little did I know then how the intensity of paleontological politics would set me up on a collision course with Phillip Tobias that almost destroyed my career.

The nineties were a golden decade for the South African search for human origins. Poised on the brink of closure at the beginning of the 1990s, due to funding constraints and Tobias's imminent retirement, the paleoanthropology department at Wits has enjoyed a resurgence of fortune. This has been due largely to the creation of a specialist funding organization, the Paleo-Anthropological Scientific Trust (PAST), which has been strongly supported by the South

African business community, to the point where 95 percent of Wits' operating costs in this sector are funded by private donors. It was this development that marked the turning point in paleoanthropology in South Africa, generating, at a critical time, the resources necessary for the rapid expansion of the science and literally saving it from extinction due to lack of funding.

The nineties also saw the most rapid transformation of South African society through a series of astounding political developments. One year after I arrived in South Africa, President F. W. de Klerk announced the release of Nelson Mandela and other imprisoned black political leaders, lifted the ban on opposition political parties, and signaled his intention to negotiate a political settlement with his enemies. Sanctions and the international academic boycott of South Africa were lifted, and the ANC formally announced an end to its armed struggle to overthrow the government. Four years of intense political negotiations accompanied by widespread political violence finally led to the country's first democratic elections in April 1994, paving the way for majority rule and Nelson Mandela's assumption to the presidency. The change of power was remarkable in that de Klerk's government willingly committed political suicide to avert a full-blown civil war on racial and ethnic lines, while Mandela used his stature to persuade a black leadership embittered by years of oppression that the path of reconciliation was preferable to that of revenge.

The 1994 elections heralded a new era. A climate conducive to a renewed search for human origins took root, attracting funding and intellectual resources to the country. International grants became available and the world scientific community began to show as much interest as had previously been shown in the traditional fossil fields of East Africa. A paleontological revolution gathered force. With new sources of funding, more fossils were found.

At the start of the nineties I'd witness a new piece of fossil hominid being found every few months, and almost every one from a single site: Sterkfontein. Now barely a week passes without an announcement of a new hominid emerging from one or another of the older established sites or a new location altogether. In 1992 we found two hominid teeth at Gladysvale, near Sterkfontein, west of Johannesburg, making this the first new early hominid site in South Africa to be discovered in 40 years. That news made world headlines. Nowadays, the novelty has worn off and new hominid fossil remains are pressed from the ground every other week, barely meriting a mention in the newspapers. This is, of course, a sign of a healthy science.

The discovery in 1992 of the first two hominid teeth from Gladysvale, attributed to Australopithecus africanus, *heralded a new era in paleoanthropology in South Africa, and made Gladysvale the first new early hominid site to be discovered in southern Africa in over 44 years.*

Within a few years of majority rule, Mandela's identified successor, Thabo Mbeki, had begun outlining his political vision of an "African renaissance," an ideology of self-assuredness that aimed to reverse the steady decline Africa had suffered despite the demise of colonialism. The significance of the South African early hominid fossil record has not been overlooked by the politicians. This was clearly evident when, in December 1998, Mbeki himself attended the announcement of the discovery of the new Sterkfontein hominid, Stw 573, after Tobias had been invited to brief the cabinet on the significance of the find. South Africa had become politically, philosophically, and scientifically trendy almost overnight, a far cry from its outcast status just a few short years before. Within a year, thanks to persistent lobbying of the United Nations Educational Social and Cultural Organization by the South African government, the Sterkfontein Valley was declared a World Heritage Site to protect the paleontological treasure chest of the country for future generations.

Without question, a great deal has changed in our view of human evolution in the years between Lucy's exhumation and today. There has been a host of new discoveries, including at least two new species that predate Lucy and may lay claim to being our oldest ancestors. There are many more fossils available to examine and interpret, and there have been advances in technology and thinking that allow us to see both the new and old fossils from different perspectives.

I'm very aware of this when I stand in front of a class at Wits with a piece of chalk in my hand, explaining what we know now about human evolution. I stress that the human family tree I draw on the blackboard is a dynamic one, that in the last 15 years there have been new discoveries in every significant period of time since our lineage parted ways with that of an ancient African ape. I use the Johanson and White model to point out how the linear chain of human evolution is fundamentally flawed and that we are forced to continually reinterpret the fossil record. When illustrating the abundance of new species, I take great care that the chalk lines I draw between the species indicate fuzzy relationships. The story today is anything but the simple family tree my professor drew on that blackboard so many years ago. It is a complex bush that is both bewildering and exciting in its twists and turns. I know one thing for sure when I finish teaching a course on human evolution today: The adventure of understanding our own evolution is just beginning.

CHAPTER 3

THE UNDISCOVERED COUNTRY

∽

Wits today is a far cry from the institution that greeted a sandy-haired 29-year-old medical graduate from Sydney, Australia, arriving in South Africa in 1923 to take over the fledgling anatomy department. Raymond Arthur Dart made his way to Johannesburg after docking by steamship in Cape Town in January of that year. He was on a mission to transform an academic backwater into a world-class place of learning. Dart was fresh from the laboratories of the great anatomist Grafton Elliot Smith at University College, London, and confessed that he was shocked when he first set eyes on the university's makeshift medical school. It was then a two-storied, L-shaped building that was nestled in the shadow of "the fort," Johannesburg's main jail.

"In those days the appearance of Johannesburg was enough to depress anyone much less sensitive than ourselves," Dart wrote. "It had such an impermanent appearance with its endless rows of red-painted, corrugated-iron roofed buildings. It seemed to have progressed little since the days of the gold rush toward the end of the last century, and one felt that if a financial slump hit the place, it would become a ghost town in a matter of days.

"My permanent staff consisted of a single preparator of material in the mortuary basement underground. The walls of the vast, high-roofed dissecting hall above this mortuary were bespattered with marks that emphasized its customary use by the students for practicing football and tennis. The zinc-covered trestle-type dissecting tables supported dried-up portions of corpses whose only coverings were hessian sheets. Our first impression left my wife, whom I had taken from her medicine studies at Cincinnati, in tears—a woman's prerogative I rather envied at that moment."

At the time of his arrival, Dart had no aspirations of becoming an anthropologist. Certainly he had no inkling that in a matter of a few years he would be thrust onto the world stage as the "father of the Taung Child," arguably the single most important anthropological discovery of the 20th century. He would claim later that he despised fossil bones, writing that he "had no sense of dedication to a search for human ancestors when coming unwillingly to South Africa.... " In fact, he continued, "I have striven ever since I was given my first piece of anatomical research work as a student in Sydney, to avoid both bones and mathematics. Circumstances thrust anthropology upon me after I had chosen to follow even more useless trails as a neurological embryologist."

Despite these protestations, Dart had neither the inclination nor the temperament to turn Taung away. Indeed, his obstinate and compulsive enthusiasm left him no option but to seize a challenge that would spark off one of the most intense and durable controversies experienced in the world of science during the past 100 years. As Phillip Tobias, who was one of his students before ultimately becoming his successor, later wrote: "So chance, peradventure or serendipity led Dart's unwilling footsteps towards anthropology. Yet he was to make one of the most seminal offerings to man's understanding of the origins of man."

If you go to Taung today it is hard to imagine that this bleak landscape was the home to such an important scientific discovery. Taung can be loosely translated from the Tswana language commonly spoken in the area to mean "place of the lion." The lime quarry where the skull was found is situated at the village of Buxton, about five hours drive from Johannesburg and eight miles from the actual town of Taung, which rests on the edge of the Kalahari Desert in the Northern Cape Province. The quarry itself is enormous, nearly two miles in length and a mile wide. It gets extraordinarily hot during summer, with temperatures rising regularly to well above 100° F. As is typical of deserts, winters can be brutally cold, nights dropping below freezing. Everything in the area, including the older houses in Buxton, is covered in a thick coat of bright white lime, hardened to the stage of being best compared to concrete. The old kilns where the lime was burned are slowly falling into disrepair, and the buildings where the mine offices were, and the original warehouses used to store equipment, stand empty. When the sun is shining brightly, one suffers a perverse kind of snow blindness from this white landscape. In the winter, the cold Kalahari winds whip the tiny particles of lime into a blinding dust storm of sorts. Despite this, in its own eerie way, it is a starkly beautiful environment. The actual site where the skull was blasted out of the limestone is today marked by a small stone pyramid, erected in 1985 on the 60th anniversary of the Taung discovery. It sits alone in the quarry below two large pinnacles of lime, which are all that remain of that section of the quarry.

The Taung Child was found by accident. Its discovery and identification were part of a remarkable chain of events. One could even venture that the Taung skull discovered Dart rather than the other way around. Like a rebel orphan who has passed through many hands, Taung was delivered to the professor's Johannesburg home on November 27, 1924, as if to claim its father. And in the same manner that a forceful and troublesome child can sometimes assume control of its parents' destiny, Taung brought Dart derision and fame, recognition and despair. By the time this little fossil was finished with Dart, it had become both his joy and his nemesis, and the plaudits he gained came at great cost to his health and personal life.

It was a hot summer in 1924, and the sound of dynamite echoed across the arid flatlands before dissipating in the desert air. These were routine explosions at the Buxton Lime Quarry outside Taung. It was the regular routine of one M. G. de Bruyn to blast out the calcified deposits of pink breccia from the dolomitic cliffs so that the lime could be used for building and gold extraction on the Witwatersrand. The monotony of de Bruyn's work was relieved from time to time by the fossilized baboon skulls that were often torn out of the lime quarry by the explosions. These skulls had been studied by Cape Town University academics since 1919 and were believed to be an extinct species. It was one such fossil that alerted Raymond Dart to the possibility that the Buxton quarries could hold other secrets of prehistory.

While de Bruyn was blasting away at the limeworks, E. G. Izod, a director of the Northern Lime Company, which owned the quarry, came across one of the skulls in the mine manager's office during a routine visit. In tracing the tale of Taung, the *Cape Times* newspaper reported:

> When Mr. Izod saw it he thought it would make an excellent paperweight, and so he took it back home with him. There it might have been to this day, but his son, Mr. "Pat" Izod, thought it would interest his student friends at the Rand University. He therefore, took it down to the University one day, and Miss Josephine Salmon [sic], a demonstrator, borrowed it to exhibit to students of Anatomy, and then Professor Dart saw it....

For his troubles, Izod senior had the honor of having the fossil baboon skull named after him by one of Dart's students, who ascribed to the specimen the label *Papio izodi*.

Dart was intrigued by what Josephine Salmons put in his hands. He was not aware that any fossil primates had been found in Africa south of Egypt and did not know of the UCT research on Taung baboon skulls. He contacted Wits University Professor of Geology Dr. R. B. Young to try and procure other specimens from Taung. Coincidentally, Dr. Young had been commissioned to investigate lime deposits not far from the Buxton Limeworks and agreed to Dart's request.

On a sweltering November morning, while Dart was soliciting more information about these fossil baboons, the quarryman de Bruyn, who had become something of an amateur fossil collector, noticed something peculiar when the dust had settled after yet another blast. Before him, exposed in a chunk of breccia, was a skull. Not the usual baboon cranium that he was familiar with but a larger specimen, one that looked more, well…human. In one account of the find, de Bruyn is said to have believed it may have been the petrified skull of a Bushman. The front part of the skull first caught his attention. It was thickly embedded in a piece of lime rock that had been propelled out of the quarry. To de Bruyn's amazement, the other half of the skull lay several yards away. The features, although half buried in the lime, were distinct. This was no baboon. De Bruyn was suitably impressed by the specimen to report it to Mr. A. E. Spiers, the mine manager, who in turn kept it in his office.

There, in the next of a string of coincidences, it was noticed by Dr. Young, who had just been briefed by Dart to watch for any unusual fossils at Taung. Young persuaded Spiers to part with the skull and had it railed, along with other specimens, back to Johannesburg. The boxes arrived on Saturday morning, November 28, at Dart's home in Melrose, Johannesburg, where frantic preparations were underway for a wedding reception.

Dart was expecting the delivery and was keen to have a quick look at the haul, despite the fact that he was to be the best man at a marriage ceremony taking place in a few hours. He could not resist wrenching open the boxes and inspecting their contents. In the first, he found little of interest; but when he opened the second, "a thrill of excitement shot through me," he recalled. "On the very top of the rock heap was what was undoubtedly an endocranial cast or mould of the interior of the skull."

By his own account, Dart was beside himself, having glimpsed the features of a creature previously unknown to science. Understandably, he was almost late for the nuptials involving one of his senior lecturers. No sooner had the wedding reception finished than Dart raced back to examine the skull, which was still encased in the hard-rock matrix formed from leeching dolomite, a substance almost as hard as concrete. He chipped away at the breccia with improvised tools, including a sharpened knitting needle belonging to his wife, Dora.

The cleanup began in earnest on Monday, December 1, 1924, when Dart and his colleague Professor Craig began to laboriously chip away the rock matrix around the skull. It took 23 days of feverish activity until the skull was exposed enough for him to properly describe it. Revealed to him was a face that has subsequently graced hundreds of textbooks and academic papers, the face of the Taung Child.

Dart's efforts eventually revealed an almost complete face, extending from the lower jaw to the forehead. The natural cast of the brain, which had been found yards away from the rest of the skull, fitted comfortably behind the bones of the face.

One of the first things about the little Taung Child that struck Dart was the shape of the teeth in relation to the size of the skull. They were unlike anything he had come across before. The mouth was humanlike, with small front teeth that met each other almost end on end instead of at the angle normally associated with apes. The tiny canines and grinders resembled those of human teeth, but the skull was too small to be that of a human. Although the upper jaw was cemented to the lower jaw, Dart noticed that all 20 milk teeth were present and the first of the permanent teeth were in a stage of eruption that could be ascribed to a six-year-old human child—or an ape a year or two younger. There was no trace of eyebrow ridges, and the forehead arched directly upward from the inward-sloping eye sockets. The face was also different. Instead of protruding like that of an ape, it was receded, contrasting markedly with the massive upper and lower jaws of the gorilla and chimpanzee. Dart estimated the skull capacity to be about 520 cubic centimeters, which was bigger than the range of all known chimpanzees, yet it was smaller than a gorilla's. The form of the brain was narrow and high, unlike those of the living apes, which are low and broad. This caused him to look more closely at the foramen magnum, the cranial opening through which the brain joins the spinal cord. This gap was at the base of the skull, not toward the rear as is the case in apes and other four-legged creatures. Dart surmised that this could only mean one thing: The creature must have walked upright.

This realization dispelled any doubt Dart may have had about the uniqueness of the find. With growing wonder, he became certain that this creature, so superbly preserved in its lime-encrusted tomb, stood somewhere between chimpanzee and human. It was certainly a completely new species. He named it *Australopithecus africanus*, which he believed filled the gap "between the most primitive of men and the most advanced of apes."

There were several elements about the find that puzzled Dart, one being its original location. The skull had emerged from a semi-desert area far from the natural habitat of apes, the closest populations of which were 1,600 miles north in what was then the Belgian Congo. Dart surmised that the Taung creatures must have been cliff dwellers, given the skull's original location in the Buxton quarry. Little did he know at the time that subsequent research would show that the Taung landscape, at the time of the child's life, was most probably subtropical.

Dart worked quickly to prepare a manuscript for scientific publication. Forty days after the little skull had found its way into his hands, he completed his paper. On Tuesday, January 6, 1925, Dart posted his findings, along with photographs of the specimen taken by a *Star* newspaper photographer, in time to catch the Cape Town mail boat to England. On Friday, January 30, the manuscript arrived on the desk of the editor of the prestigious science journal *Nature*.

Raymond Dart examines the Taung Child skull in 1925. This find would singularly show that the roots of humanity lay in Africa and not in Asia or Europe, although it would take more than two and a half decades for its significance to be recognized.

The arrival of this challenging bundle of information put the editorial board of *Nature* in a quandary. Unmistakably, this was a species new to science, yet the accepted protocol of the time was that a specimen should be studied for five years before any findings were published. Dart had taken barely a month and a half. Furthermore, instead of confining himself to a preliminary description of the fossil, Dart leaped to a series of breathtaking conclusions about his new species:

> They possessed to a degree unappreciated by the living anthropoids the use of their hands and ears and the consequent faculty of associating with the color, form, and general appearance of objects, their weight, texture, resilience and flexibility, as well as the significance of sounds emitted by them. In other words, their eyes saw, their ears heard, and

their hands handled objects with greater meaning and to fuller purpose than the corresponding organs in recent apes. They had laid down the foundations of that discriminative knowledge of the appearance, feeling, and sound of things that was a necessary milestone in the acquisition of articulate speech.

Dart had also released his paper to the *Star* with the instruction that it should not be published until the article had appeared in *Nature*. The *Star*, however, decided to break the embargo after sensing that *Nature* was in a dither, and on Tuesday, February 3, the day before Dart's 32nd birthday, it published its world scoop, splashing Taung across its front page.

In fact, *Nature* had decided to publish Dart's findings in its February 7 issue. Proofs of the manuscript had been circulated for comment to four eminent British prehistorians—Sir Arthur Keith, Sir Grafton Elliot Smith, Sir Andrew Smith Woodward, and W.L.H. Duckworth. Headline news tends to gallop, however, and the scientific establishment was taken by surprise. The same morning as the *Star* published the Taung story, Sir Arthur Keith received the proofs from *Nature*. After reading through Dart's article and examining the illustrations, he concluded that the skull was that of an anthropoid (ape). He wrote in his diary: "On the night of the 3rd reporters came in troops in consequence of a cable from South Africa. I kept quiet and left the talking to Elliot Smith."

Although Sir Grafton Elliot Smith, Dart's former teacher, was intrigued by the find, he reserved his opinion, pending the chance to study the skull itself. Sir Andrew Smith Woodward dismissed the find as meaningless, while W.L.H. Duckworth, another of Dart's mentors, believed the specimen to be most closely related to a gorilla. These comments, along with Sir Arthur Keith's negative reaction, were published by *Nature* in its next edition, a week later.

Dart's seemingly undue haste in publishing his findings was to count against him for many years as many scientists felt due process had not been observed. Dart also had something of a reputation. According to the distinguished anatomist Sir Wilfred Le Gros Clark, Dart was known as a man who "might be inclined too hastily to arrive at conclusions on too little evidence."

In an interview with the *Star* published on February 7, Dart gave his own summary of the discovery:

> The geological record of the different species of man has been rendered fairly perfect; where geological evidence has been lacking is in specimens of that phase of pre-human existence between the most primitive of men and the most advanced of apes. This gap is now filled by Australopithecus africanus. The Taungs individual, in brief, was not a human being, and yet he was a much more intelligent being than the gorilla or the chimpanzee which are the highest of the living apes. He was unable to talk but his brain was advanced in the direction required in an ancestor whose offspring were to attain ultimately the power of communicating with their fellows by the symbolism of speech. He is therefore to be regarded, not as an ape-like man but rather as a man-like ape.

Dart's critics have often been vilified in history for their hesitancy in accepting the little child outright, but to be fair, many recognized the importance of the find but shirked from the conclusions. In an interview with the *Rand Daily Mail*'s London correspondent, published on the same day as the *Star* interview with Dart, Sir Arthur Keith, later often portrayed as Dart's archrival, was quite sympathetic. Although he still held the view that Taung was not ancestral to modern humans:

> Professor Dart deserves great credit. He has certainly lived up to the opinion we formed of him here (in London). I think he has done very well indeed. After all, his discovery should not come in the nature of a shock. It is rather in keeping with developments that have been going on in South Africa for some years. What do the geologists say regarding this discovery? Am I right in supposing Dart's find was made in a filled-up cave that had been obliterated in the course of time? We want to know when Dart's ape-man was living. The geological evidence has a tremendous amount to do with the meaning of Dart's discovery. My opinion is that perhaps it is not very old geologically. Then we have to consider what sort of beast it is that Dart has got hold of. I certainly agree with him in all of his main contentions. What he has discovered appears to be an unknown form of man-like ape. The photographs

impress me with the similarity to the gorilla and the chimpanzee, and the impression I form is that it is really nearer akin to them. But I also agree with Dart that it is more human than either.

The Taung discovery took place against the backdrop of the Piltdown controversy. This was one of the most daring frauds of 20th-century science and one that would fundamentally affect the course of the search for human origins for almost 40 years. In 1908, workers digging a gravel pit close to Piltdown Common in Sussex, England, recovered a number of skull fragments. These were handed over to an amateur paleontologist, Charles Dawson, who in 1912 presented these fossil relics to British Natural History Museum geologist Arthur Smith Woodward. Smith Woodward and Dawson, accompanied by the theological mystic and Jesuit paleontologist Pierre Teilhard de Chardin, conducted further excavations of the Piltdown pits and found more pieces of a skull together with part of the right half of an apelike lower jaw.

Late in 1912, Smith Woodward pieced together the fragments and unveiled what he called *Eonthropus dawsoni* ("Dawson's dawn man") at a meeting of the Royal Geological Society. Sir Grafton Elliot Smith analyzed the cranium and found it to be "quite simian." He soon convinced himself that Piltdown was indeed the ancestor of modern humans, a point seized upon by the *Illustrated London News* of December 28, 1912, which proclaimed:

> A discovery of supreme importance to all who are interested in the history of the human race was announced at the Geological Society when Mr Charles Dawson of Lewes and Dr A Smith Woodward, the Keeper of the Geological Department of the British Museum, displayed to an eager audience part of the jaw and a portion of the skull of the most ancient inhabitant of England, if not in Europe.

Many of those who disagreed with this conclusion still accepted the Piltdown skull's authenticity. Sir Arthur Keith, for instance, accepted Piltdown Man's credentials but believed it to be a Neandertal representing a dead-end in human evolution. Others, though, like the American mammologist Gerrit S. Miller, believed that the Piltdown skull combined a human cranium and a chimpanzee jaw, an observation that proved not to be too far off the mark when the hoax was eventually exposed during the 1950s.

Other jaw fragments from the Piltdown pit were recovered in 1913, and analysis of the skull continued into the 1920s. By this time it had become conventional wisdom in the form of a Pre-*sapiens* theory: At some remote point, probably in the Pliocene, a split occurred in the human lineage, with one branch leading, via Piltdown, to modern humans, the other leading to the doomed Neandertals. This view was challenged by the Czech-born American physical anthropologist Alex Hrdlicka, who insisted that the skull and jaw were from different animals and the skull was far younger. Hrdlicka was right, but his views were advanced to support his belief that Neandertals were our ancestors.

Piltdown was not a scientific mistake. It was a hoax. It emerged in 1953 that the cranial fragments of a modern human had been stained to make them appear older and that the jaw was indeed that of an ape with teeth that had been filed down. They had been planted in the pit, along with stone tools and mammal fossils, for the paleontologists to find. To this day, we do not know with absolute certainty who perpetuated the Piltdown hoax. Dawson had been the most often accused, and evidence has emerged that conclusively links him to the act. Many feel, however, that the sophistication of the fraud required a greater scientific knowledge than that possessed by Dawson, and fingers have been pointed at almost every person who came in contact with the specimen, including Sir Arthur Keith himself.

Piltdown remains a wonderful whodunit story. At the time, however, it distracted the scientific establishment from the importance of Taung and the implications that this little skull held for the ancestry of humankind. Dr. Henry Fairfield Osborn, director of the American Museum of Natural History, for instance, was a man who accepted the principle of evolution. But he felt passionately that humanity had developed over a much longer time period than that suggested by Dart. "The most welcome gift from anthropology to humanity," he wrote, "will be the banishment of the myth and the bogie of ape-man ancestry and the substitution of a long line of ancestors of our own at the dividing point which separates the terrestrial from the arboreal line of primates."

Dart was also criticized on more spurious grounds. He was condemned, for instance, for mixing Latin (*australo*) with Latinized Greek (*pithecus*), prompting his colleague Robert Broom to remark snidely that in the face of one of science's most important discoveries, certain academics got into a froth about grammar. But from a scientific view, even Dart's sympathizers had to concede that the youthfulness of the Taung specimen was a disturbing factor: Who knew what it might have looked like as an adult? It was unacceptable, it was argued, to use an immature specimen to diagnose a new species.

Furthermore, Dart had not even attempted to put a date on the age of the fossilized skull. This, along with the fact that it had not been studied in its context, dented his credibility significantly. Dart compounded the criticism he faced by his reluctance to allow anyone else to personally study the skull. The first time many scientists had an opportunity to actually see the specimen was at the British Empire Exhibition in the summer of 1925, where they had to rub shoulders with the crowds to get a view. As Ian Tattersal of the American Museum of Natural History has pointed out: "The British paleoanthropological establishment, forced to peer at the specimen through glass while being jostled by the passing hoi polloi, was not amused."

Keith in particular waded into Dart after the British Empire Exhibition, writing in the July 4, 1925, edition of *Nature*, "An examination of the casts exhibited at Wembley will satisfy zoologists that [Dart's] claim is preposterous. The skull is that of a young anthropoid ape.... The Taungs ape is much too late in the scale of time to have any place in man's ancestry."

It was only in 1931 that the world's leading scientists had the opportunity to examine the skull. Dart took it to London where he was invited to take part in an international anthropological congress to defend his theories. Taung, however, was overshadowed by the discovery of hominid remains in a cave in China. A fossil hominid tooth found in a Chinese drugstore was traced to a cave near the village of Choukoutien, where the bones ascribed to Peking Man (eventually classified as *Homo erectus*) were uncovered, thus giving China a claim to being humanity's ancestral source. Peking Man became all the rage, and at the conference, Dart was preceded by a presentation of the findings from China. Dart then stood up to churn out his rather tired defense of Taung, which contained nothing he had not already said. He was not enthusiastically received.

Dart conceded defeat and lapsed into a deep depression. He was taken off to dinner by some colleagues intent on trying to cheer him up while his wife, Dora, caught a taxi back to their hotel, clutching the box that contained the precious skull. An incredible thing then happened. She forgot the box on the backseat of the cab. There the Taung skull remained for the entire night, being ferried around with numerous fares until the taxi driver found it and handed it over to the police. God knows what went through their minds when they opened the parcel and were confronted by the skull. By this time, Dart, having returned from dinner and beside himself with worry at the loss of his "child," had contacted the police himself. Scientist and specimen were eventually reunited. Dart returned to South Africa, made abject by the ongoing criticism he faced from the British scientific establishment, even though many now had the opportunity to personally examine the skull.

In retrospect, the main problem with Taung was probably one of prejudice. Piltdown had created a mindset, particularly in Britain, as to where the ancestors of modern humans had developed. Taung had an ape brain with a human jaw. If you had to design something that would be the opposite of Piltdown, Taung would be it.

Africa, the "dark continent," was not where the Victorian establishment expected to find the clues to human origins. The prevailing view, influenced mostly by the discoveries of Java Man by Eugene Dubois between 1889 and 1895, was that humankind had evolved either in Europe or in Asia. Dubois, a Dutch surgeon, had discovered a fossilized skullcap, thighbone, and other fragments of what he called *Pithecanthropus erectus* ("the upright ape-man"), although it was subsequently reclassified as *Homo erectus*. The discovery of Peking Man during the twenties reinforced the view that humanity had surfaced in the Far East.

As Phillip Tobias has pointed out, Dart's problems were as much political as scientific: "The preconceived theories of man's origin were heavily weighted against the suggestion of the little-known young anatomist from the unlikely end of Africa. Not only was Dart's discovery made in the wrong part of the world; it was the wrong kind of creature." While many scientists moderated their criticism of Dart in the more dignified tomes of academia, the popular press, particularly in Britain, was less subtle in voicing its opposition to the theory both in editorials and letters. One paper suggested that Dart be placed in a home for the feebleminded, while the deeply religious went even further: A Frenchman predicted that Dart would "roast in the quenchless fires of hell." A Dane warned that Dart had signed his "perdition-warrant" by having the impertinence to account for the origin of man, while another critic accused Dart of being a priest of Baal. Under the title "Hammer and Taungs," a letter in the *London Sunday Times* was signed "Yours respectfully, a Plain but Sane Woman." Dart was labeled a traitor to his Creator for making himself "the active agent of Satan and his ready tool." He had clearly touched on a scientific nerve, one that is still surprisingly raw today, with many people still refusing to believe that we have descended from a line of ancestral apes. These teachings were anathema to fundamentalist Christianity. Dart's name was invoked in the passing of laws in several American states that outlawed the teaching of evolution, while the Taung skull was referred to derisively in a parliamentary debate in Cape Town. The conservative British magazine, the *Spectator*, invited readers to contribute suitable epitaphs for *africanus*, offering a prize for the best backhanders.

Not everyone, however, gave Taung a frosty reception. Among the messages of support that Dart received was from the then Prime Minister Jan Smuts, who in his capacity as the president of the South African Association for the Advancement of Science, wrote to "express the hope that many further triumphs await you." Jan Hofmeyer, the Administrator of the old Transvaal, and a former principal at Wits, wrote that the discovery was "one of the things that makes one [occasionally] regret having left the University."

Dart's greatest ally came from an unlikely source, the rather eccentric Robert Broom, a Scottish midwife who's passion was the study of Karoo mammal-like reptiles. Unconventional and arrogant, Broom was never a man short of enemies, although he managed to outlive most of them.

Born in 1866 and raised in conditions of dire poverty, Broom overcame the impediments of a 19th-century working-class pedigree and only four years of formal education to eventually graduate with honors from Glasgow University in 1889. Ever contemptuous of people in positions of authority or wealth, Broom traveled the world, practicing as a locum in all manner of isolated areas until he came to South Africa from Australia in 1897. A restless man, he moved about the towns of the rural northern Cape, followed by his long-suffering wife, a Scottish servant girl he had met during his studies. He capitalized on his lifelong obsession for fossils by becoming a world authority on the mammal-like reptiles of the prehistoric Karoo and attracted the interest of the world's leading museums. It was not long, however, before he drew animosity from some of his colleagues for allegedly supplementing the income from his medical practice by selling Karoo fossils to scientific institutions abroad. At one stage, he was barred from entering the South African Museum after being accused of stealing fossils from other collections.

Broom also had the reputation as a womanizer. At the age of 74, he said: "I am only reminded of age when I climb a hill or catch a bus. My heart still goes pit-a-pat when I see a pretty girl and we still have quite a lot at Pretoria. They say sex-attraction still continues in Paradise—what a time I am looking forward to."

No one could deny, however, that Broom was a prodigious worker. By 1925, when the Taung Child's existence was revealed to the world, the 58-year-old Broom had authored 250 publications, and named some 70 new genera and almost 200 new species of reptiles. He also displayed a remarkable sense of energy and vigor, as described by Roy Terry, who also wrote that Broom had been a man in a hurry during his whole life: "Until his death he scarcely ever walked, preferring a brisk trot bordering on a run. Even as a septuagenarian he could climb hillsides faster than men half his age. His speech was so rapid that shorthand writers found it difficult to keep pace. No one ever saw signs of fatigue or heard him complain that he was tired and he could exhaust most people both mentally and physically."

Broom was practicing as a doctor in the small Karoo town of Douglas when news of Taung reached him. He immediately wrote to Dart congratulating him on the find, and within two weeks of the revelations made his way to Wits Medical School. There, according to Dart, Broom burst in unannounced and "ignoring me and my staff, he strode over to the bench on which the skull reposed and dropped on his knees 'in adoration of our ancestor' as he put it." Dart granted his request to make a personal study of the skull, and Broom spent the weekend of February 21, 1925, with him. Robert Broom came out fighting for Dart almost right away. He lost no time in writing to *Nature* in support of Dart and was published in the April 18, 1925, edition, saying he was convinced that in Taung "we have a connecting link between the higher apes and one of the lowest human types."

It soon became apparent that Broom was not in a position to do much for Taung other than to express his support. He did not have the credibility to swing academic opinion, because the reputation he had built up in describing the Karoo mammal-like reptiles was dented by institutional hostility toward him. Besides that, his commitment to Dart and the little skull was not going to put food on the table. The truth of it was that Broom was barely making ends meet, though he was working as a doctor in a number of small sheep-farming towns in the Karoo. The onset of the Great Depression in the early thirties and the collapse of the South African wool industry compounded his plight as even his more wealthy patients suffered a reversal in fortunes.

The situation became so desperate that, in 1933, Broom could not afford the train ticket to make a speech to the South African Association for the Advancement of Science, despite the fact that he was the organization's president. When Dart learned of Broom's circumstances, he appealed to newly elected Prime Minister Jan Smuts to use his influence to assist the ailing scientist. Smuts, an ardent supporter of natural history, persuaded the Transvaal Museum in Pretoria to offer him a post. Grudgingly, the museum gave him a temporary position on the understanding that he was forbidden to collect fossils for himself or anybody else.

Circumstances then began to look more favorably upon the elderly Scottish doctor, and in 1934 he was appointed the museum's Curator of Vertebrate Paleontology and Physical Anthropology. He finally found himself in a position to commit some resources to supporting Dart's quest. Mindful that the most valid criticism of Dart's find was the fact that Taung was a juvenile, Broom became certain that what he needed to do was find an adult australopithecine. Broom centered his search around the West Rand near Johannesburg.

Then, in 1936, Sterkfontein came to light. This impressive set of caves is part of an extensive network of subterranean cavities beneath the dolomitic hills outside Krugersdorp, a satellite gold-mining town some 40 miles west of Johannesburg. Sterkfontein's existence had been known for some time, but it had somehow escaped serious scientific inquiry. Fossils had first been found there in 1895 by a group of boys from Marist Brothers College in Johannesburg, and then subsequently by prospectors and miners. Several references were made to the amount of mammalian fossil material at the site by the South African Geological Society in 1897, and fortunately during that year, Mr. David Draper intervened to prevent the destruction of the cave. He persuaded the company that owned the mineral rights to preserve the main cavern because of its impressive stalactite and stalagmite formations and underground lake. Had he not done so, the caves would probably have been dynamited by lime workers, and South Africa's most famous fossil site would today be a mere hole in the ground.

Although the caves themselves were saved—and became something of a tourist attraction during the early years of the 20th century—the quarrying of lime continued in the immediate vicinity. In 1935 one of Dart's assistants, Trevor Jones, visited Sterkfontein and was struck by the number of baboon fossils he found in the lime breccia. In July 1936, two of Dart's students visited Broom with a fossil brain cast from Sterkfontein and suggested that he pay more attention to the site. Ironically, Broom came across a little guidebook to the area written by the owner of Sterkfontein, R. M. Cooper, who sold fossils to tourists from the tearoom next to the caves. It had on its front cover, "Come to Sterkfontein and find the missing link." In a manner of speaking, that is exactly what Broom did. Broom, who by now had turned 70, felt he was getting closer to fulfilling his vow to find an adult specimen of *A. africanus* before he died. He visited Sterkfontein for the first time on August 9, 1936. By coincidence he found that the quarryman was one G. W. Barlow, who had worked at Taung when de Bruyn had discovered the famous skull. Broom asked him whether anything like that had been found at Sterkfontein and recorded the quarryman's response: "He said in an off-hand manner that he rather thought there had been. So I asked him to keep a sharp look out; which he promised to do." Nonetheless Broom was appalled. "It's sad to think that for nearly 40 years no scientist ever paid the slightest attention to these caves; and probably some dozens of skulls of ape-men and all the bones of their skeletons were burnt in lime-kilns."

Broom later recounted that on August 17, when he again visited Sterkfontein, "Barlow handed me the blasted-out natural brain cast of an anthropoid and said, 'Is this what you're after?' I replied 'That's what I'm after.'" For hours Broom sifted through the blasted breccia in the hope of getting other parts, but with no success. The next day he returned with a number of assistants and found the base of the skull and a number of bone fragments, including a jawbone and teeth. He would eventually find an almost complete pelvis with an upper bone and shoulder blade of what looked suspiciously like that of his holy grail, an adult *africanus*. But Broom found it difficult to compare his new find to the Taung skull.

"All we could compare were the young brain cast with the adult brain cast, and the unworn first upper molar of the child and the worn molar of the adult. Undoubtedly the two forms were allied, but they seemed to be at least specifically distinct so I called the Sterkfontein skull *Australopithecus transvaalensis*." On further examination, Broom later decided to place the specimens in a new genus altogether, calling it *Plesianthropus transvaalensis*. He wrote to *Nature* saying this find confirmed Dart's original thesis as to the existence of an intermediate species between man and ape that had lived in Africa. His views were politely but unenthusiastically received, because scientific opinion still swayed in favor of Asian origins.

While Broom proved quite resilient to the hostility toward Taung, Dart took it all personally. He was overwhelmed by the criticism he faced and gradually retreated into an academic exile, concentrating his efforts largely on building the anatomy department, where he also developed the reputation of being something of an eccentric. He was regarded as a punishing taskmaster whose notorious temper often reduced those around him to tears. One of his colleagues remarked: "Students either hated him or adored him. Few remained unaffected, and throughout his life Dart has always provoked strong emotional response in the people he meets."

But he also developed a repertoire of lecture-hall tricks that thrilled and astounded his students. He used the time he spent in 1930 studying gorillas in the Ituri Forest in the then Belgian Congo to demonstrate the brachiation form of primate locomotion by leaping up and grasping the hot-water pipes in the classroom, swinging from one to the other. Or he would be down on all fours on the floor crawling around like a crocodile to illustrate how reptiles moved around.

Dart grew increasingly emotional in the pre-World War II years and became prone to outbursts that alternatively endeared him to and alienated him from his colleagues. When asked about his bouts of weeping when pleading a cause, Dart once exclaimed: "I have little control over my sympathetic nervous system."

Nonetheless, the strain imposed by Taung is blamed by some for the breakup of his marriage to his first wife, Dora. Dart later married Marjorie Frew, a librarian at Wits, but continued to bury himself in his work, arriving early in the morning at the medical school and leaving late at night. Eventually, in 1943, he suffered a nervous breakdown and was forced to take a year off work. He adopted bricklaying as a hobby to soothe his shattered nerves.

Dart's heritage—as well as those who succeeded him—is evident in the hundreds of bone fragments that are neatly stored in the large fossil hominid walk-in safe at the University of the Witwatersrand. Some of the most precious fossils on the planet are stored here, among them one kept in a small wooden box that rests on the top shelf just to the right of the door. Inside this unassuming container rests the Taung skull, the first fossil human ancestor ever discovered in Africa and one of the most important finds in the history of paleoanthropology.

The whole skull is about the size of a grapefruit, and although half the brain cast is missing, it has an aura of perfection about it. When held to the light, the surface is transformed into a beautiful field of calcium carbonate crystals that sparkle like diamonds. It is as if nature has compensated for the missing pieces by creating this little crystal wonderland through the process of fossilization. As the tiny child's skull lay on its side, sediments seeped through the foramen magnum, the hole at the base of the head, filling half the skull. The rest became a miniature cave complete with stalagmites and stalactites, with only the tiny stalagmites remaining. It is also probably singly the most beautiful fossil of an early human ever discovered. Besides the crystal field adorning the brain cast, the surface of the fossil bone of the face and jaw is a soft cream color, while the endocast of the brain is preserved in a deep ocher-colored limestone.

During the late thirties, Dart became an intensely private man stung by the personal attacks he had suffered over Taung. Gradually Broom assumed the mantle as the champion of *africanus* and became determined to find a well-preserved adult specimen of the ape-man. Broom continued to focus his attention on the Sterkfontein area in his hunt for the elusive fossils. His use of dynamite caused the consternation of many of his colleagues who, quite rightly, were concerned that this method of fossil extraction could do more damage than good. Broom was undeterred by their criticisms and, dressed in his trademark black suit, could be seen crawling around on his hands and knees as the dust cleared, "sifting earth, peering into obscure corners and inspecting tiny fragments."

On June 8, 1938, the quarryman Barlow contacted Broom and said that he had found something special; it appeared to be part of an australopithecine palate and a first molar. Broom immediately purchased it from him, but was puzzled because the matrix in which the fossil was set was different from the rock around Sterkfontein. What really seized his interest was the shape of the jaw. It was far more robust than the kind of specimen he was looking for. Barlow insisted that it was from Sterkfontein, but Broom did not believe him. He returned a few days later when the quarryman was away and showed the specimen to the workers at the quarry. They had never seen that fossil before. Broom confronted Barlow a few days later saying it was vital that he know exactly where the palate had been discovered. Barlow admitted then that it was not from Sterkfontein but had been dug up by a schoolboy, Gert Terreblanche, on the neighboring farm of Kromdraai, just over a mile away. The 15-year-old boy acted as a guide to visitors to the caves on Sundays and apparently knew the area very well.

Broom immediately went to the farm but found that Gert was at school. He persuaded the boy's sister to take him to the hillside where the fossils had been found. After fossicking in the dust, Broom found a fossil tooth, which was enough evidence to persuade him to walk straight to the young Terreblanche's school to find him.

At the school, Broom explained his mission to the principal. Broom later wrote that when Gert was duly summoned, he "drew from his trouser pocket four of the most wonderful teeth ever seen in the world's history." Broom persuaded him to surrender them for a shilling apiece. Under the Scottish doctor's gentle interrogation, he admitted prizing the teeth out of what appeared to be a jawbone embedded in the rocks, and said he had some more pieces hidden away. As there was still another hour and a half of school time left, Broom suggested to the headmaster that he lecture the 4 teachers and 120 children at the school on the importance of fossils and how caves were formed. Broom as always was dressed in his formal black suit and starched wing collar, giving him an appearance of grave authority, and had no difficulty enthralling his young audience about how to look for fossils. Once the class was over, Gert took Broom to the fossil site, where Broom found the skull from which the teeth had come. Although the skull had been smashed by the schoolboy's crude dental extraction, there were sufficient fragments to allow a reconstruction. This material plus another tooth that Gert had stashed away were handed over in exchange for five chocolate bars.

Subsequently the whole area was thoroughly searched and the topsoil sieved by members of the staff of the Transvaal Museum. The result revealed major portions of another skull that could be adequately reconstructed—but it varied considerably from earlier finds. It was altogether more massive, particularly the jaws, which carried unusually large molars and were equipped with bony ridges to anchor extremely powerful muscles. Broom labeled him *Paranthropus robustus*—a robust creature akin to man.

The discovery of the Kromdraai fossil (labeled TM 1517) ranks high in the study of human evolution and provided Broom with the decisive break that began turning world opinion back in Dart's favor. This was due partly to the coincidental arrival in South Africa of two of the most influential American scientists, William K. Gregory and Milo Hellman, who had come to examine the Taung skull and the Sterkfontein specimens. They concluded that these were "in both a structural and a genetic sense the conservative cousins of the contemporary human branch." Gregory told a meeting of the Associated Scientific and Technical Societies of South Africa on July 20, 1938, that "the whole world is indebted to these two men for their discoveries, which have reached the climax of more than a century of research on that great problem, the origin and physical structure of man." This was the first glimmer of a shifting tide of scientific opinion toward Dart and Broom, but it would take another decade before final vindication.

Broom's search for a well-preserved adult of the Taung Child was painfully slow, interrupted increasingly by the war clouds that gathered over Europe, where hostilities eventually broke out in 1939, and when South Africa went into the Second World War on the side of the Allies.

But even World War II could not bring a halt to Broom's enthusiastic search. Broom began further excavations at Kromdraai in 1941, but his mood gradually turned to disappointment as the sterile breccia revealed few fossils of significance with the exception of a young robust australopithecine jaw. Broom closed down the site and focused on a monographic treatment of his finds. He worked on this during the war years and published *The South African Fossil Ape-Men: The Australopithecinae* on January 31, 1946, shortly before he turned 80. His conclusions confirmed Dart's original hypothesis that the australopithecines had walked upright, and although they had small brains and ape-like faces, their teeth were humanlike. He believed they had the ability to use tools and probably lived in open country. Broom received the U.S. National Academy of Sciences Award for the most important biological book of the year. Like Dart before him, however, Broom was criticized for being too hasty to arrive at his conclusions, which were considered by Wilfred Le Gros Clark to be overambitious.

Dart, in the meantime, was still nursing the wounds he had received in his fight to get Taung recognized. His confidence must have been boosted by the recognition that Broom had received for his monograph, and his state of mind must have been lifted again later that year when Makapansgat was drawn to his attention. This was a set of caves just outside Potgietersrus in the Northern Province. The cave had been declared a National Monument ten years previously in commemoration of the subjugation in 1854 of Chief Makhapane's tribe by the Boers, South African farmers of Dutch origin. The story of Makhapane is a tragic one, an illustration of the violence between black and white South Africans that still echoes through that country's national psyche today. In retaliation for the murder of a party of 14 white men, women, and children by Makhapane's warriors, the Boers, who mustered a fighting force to teach Makhapane a lesson, made a clarion call to arms. His tribespeople, seeing trouble coming, retreated into the enormous cave, 2,000 feet long and 500 feet wide. Here they were besieged by the Boers, who decided to starve them into surrender by shoring up the mouth of the cave with thorn trees and rocks. For almost a month, Makhapane's supporters held out, engaging in occasional skirmishes with the Boers, but ultimately their spirit was broken. Of the tribespeople who had fled into the cave, only 175 emerged into the harsh sunlight to give themselves up, leaving over 1,500 bodies in the stench-filled darkness behind them. Makhapane's name was given to the caves (*gat* in Afrikaans), and the bones of his supporters lay for many years in the recesses of the cave system, and can still be found today.

Makapansgat was a rich source of much older bone material, preserved, as in the case of Taung and Sterkfontein, in the dolomitic lime that characterizes the geology of the area. Dart was made aware of the prehistoric possibilities of the site when he was sent fossil bones by a local schoolteacher named Wilfred Eitzman, who found the bones there in 1925. It was only in the early forties, when a visitor to the site found the fossil skull of an extinct baboon species, that the site aroused interest at Wits University.

One of Dart's students, the young Philip Tobias, examined the baboon fossil and persuaded Dart that the site had potential.

In 1945 Dart sent 30 of his anatomy students, under the guidance of Tobias, to have a closer look. More baboon fossils were recovered from what was called the Cave of Hearths, which encouraged Dart, still suffering from the effects of his nervous breakdown two years earlier, to raise funds for a proper excavation of Makapansgat. Fortunately, a patron stepped forward in the form of Dr. Bernard Price, a wealthy Scottish electrical engineer, who offered a donation of a thousand British pounds sterling a year for research to be conducted at Makapansgat. This paid for the 1947 expedition under James Kitching, who began a large-scale systematic search through the caves. Kitching, who eventually became well known for his extraordinary ability to find Karoo fossils, soon discovered the first remnants of a gracile australopithecine in the form of a small piece of skullcap.

Dart, whose health was deteriorating, nonetheless allocated the finds to a new species *Australopithecus prometheus*, named after the Greek hero who stole fire from the gods. This was because he believed the blackened condition of many of the Makapansgat bones showed that these creatures had used fire. It later turned out that the blackening was due to the deposition of manganese on the bones and not burning. The small group of fossils from Makapansgat would later be reclassified as *Astralopithecus africanus*.

Most important to scientific progress was that at both Makapansgat and Sterkfontein substantial pieces of pelvic bones were found. They established beyond doubt what Dart had postulated from the Taung skull more than 20 years before—that australopithecines walked upright. The evidence backing Dart and Broom's claims about *africanus'* prehuman ancestry was beginning to look impressive.

The first Pan African Prehistory Conference held in Nairobi in January 1947 marked the coming of age of the South African search for human origins. This gathering, organized primarily by Louis and Mary Leakey, was also significant in that it was the first African scientific initiative to challenge the Euro-American hegemony in the search for human origins. Indeed, given the postwar tensions that were still pervasive in Europe, particularly in Paris, which was then the center of prehistory studies, Africa was an ideal, neutral location for such an event. Louis Leakey recalls in his memoirs, *By the Evidence,* that many French scientists were very bitter toward the leading prehistorian of the time, the Abbé Henri Breuil, who had spent the greater part of the war in South Africa at the invitation of Prime Minister General Jan Smuts. Those Frenchmen who remained behind when the Germans overran France regarded the Abbé as a sellout for turning his back on the war effort. Leakey had a hard time persuading the French that the Abbé Breuil should be the first president of the Nairobi congress and that the war years he had spent in Africa had contributed to the science.

General Smuts himself took a personal interest in the proceedings and made a military aircraft available to transport scientists to Kenya from southern Africa, including the Portuguese colonies of Mozambique and Angola. In fact Smuts, who had become an ardent student of prehistory under the guidance of a friend, archaeologist C. van Riet Lowe, sent with his delegation an invitation to scientists for a follow-up conference to be held in 1948 in South Africa. This was not to be, as Smuts was unseated by the politics of Afrikaner nationalism, a movement that viewed the study of human evolution with creationist hostility.

From a South African perspective, the Nairobi congress was extremely important in showcasing all the research that had gone into the australopithecines. After the official opening by the British governor of Kenya, Sir Gilbert Rennie, Dart, and Broom were given the opportunity to describe the specimens from Taung, Sterkfontein, and Kromdraai. Le Gros Clark then led a general debate in which he threw his weight behind the South Africans and, through forceful argument, convinced many of the skeptics. After the Nairobi plenary, the 81 delegates then visited a number of East African sites, including Olorgesailie, where much to the delight of his detractors, the Abbé Breuil's braces broke during his speech, and he had to use one hand to hold up his trousers while he talked.

Broom was in his element during the conference and impressed all with his boundless enthusiasm. During an excursion to a rock art site near Kisese, he ignored a plea by Leakey to refrain from the two-mile walk to the cave paintings in the blazing summer sun. Leakey wrote: "I shall never forget the sight of Robert Broom—then almost 80 years old—wearing, as always, a dark suit, wing collar, and butterfly tie, negotiating the last steep stretch in the heat of the day. It was indeed an amazing feat for a man of his age in such unsuitable clothing." The next day, on their return to Nairobi, the delegates found the main road flooded and had to wait until the river subsided before they could cross. Taking off their socks and shoes, they were "led magnificently by Robert Broom, who rolled his black trousers up to his knees and strode ahead of everybody."

The First Pan African Prehistory Congress was a great success, wrote Leakey. "It resulted in a general feeling that Darwin's prophecy that Africa would prove to be the birthplace of mankind was correct. In particular, Le Gros Clark, as the leading anatomist at the congress, made a major contribution to this end. Since he had just come from South Africa, where he had seen the australopithicene specimens at Pretoria and Johannesburg and discussed them [with Dart and Broom], he could speak of them with the authority of firsthand knowledge. At the congress he made it abundantly clear that there was no longer any doubt in his mind that these South African manlike fossils were hominid, not pongid [apelike], and that they were more closely related to man than they were to any ape."

Broom returned to South Africa feeling vindicated and resumed his fossil hunting in the dolomitic hills west of Johannesburg. His eccentricity had even his supporters concerned, however, particularly about his ability to catalog and accurately record what he was blasting out of the West Rand sites. His archaeologist colleague H.B.S. Cooke recalled that "one day [in 1947] Lawrence Wells and I were in the Transvaal Museum and asked to see the Kromdraai skull. Broom fossicked in a drawer and pulled out the facial part, then rushed off down the corridor, with us on his trail, and into the laboratory of Vivien Fitzsimmons. There he pushed aside two jars of snakes and said: 'Here's the other piece, I knew it must be there.' We were very perturbed that all this valuable material was unlabelled and uncatalogued and that only Broom knew what was what. I happened to know General Smuts personally and we persuaded him to back the appointment of an assistant to ensure that proper records were established. By great good fortune the appointee was John Robinson, a graduate of the University of Stellenbosch, and soon he began to earn a name for himself in his own right."

Despite his successes, Broom became more controversial, particularly in his liberal use of dynamite to extract fossils from limestone, a practice that set him on a collision course with the National Monuments Council. Many archaeologists felt, with justification, that explosives destroyed the context of the fossils, making it impossible to try and date them. Broom's response was that the rock-hard nature of the sediments left him little choice but to employ such drastic measures. The council disagreed and banned him from further work unless a competent field geologist accompanied him. Broom, who'd served as a professor of geology at Victoria College in Stellenbosch, was outraged and appealed to Prime Minister Smuts among others for support. The council backed down, and a defiant Broom continued blasting at Sterkfontein. Within a few days his efforts paid off. On April 18, 1947, he and his assistant John T. Robinson found the near-perfect cranium of an adult australopithecine, which he named *Plesianthropus* (near man) *transvaalensis*, still known today as "Mrs. Ples" despite being reclassified as *A. africanus*.

"I have seen many interesting sights in my long life but this was the most thrilling in my experience," he wrote after a lucky blast exposed the almost complete skull of Mrs. Ples in two parts, revealing a crystal-lined brain cavity. After a few weeks of laboratory work in which the skull was cleaned and prepared, Broom and Robinson concluded that the skull was that of a middle-aged female, hence the label Mrs. Ples. On August 1, Broom and Robinson blasted out a slab of breccia that contained a partial thigh bone, several vertebrae, and a more or less intact pelvis of an australopithecine (labeled Sts 14). This convinced them completely that Dart had been right in his assessment of the Taung Child, that the ape-men could walk upright. By the end of 1947, Broom had amassed other specimens, including a lower jaw with teeth.

Mrs. Ples, or Sts 5, "represented a vital turning point in the broader acceptance of South African australopithecines as hominids. With an adult cranium like Sts 5, the case could no longer be made that the Taung Child was simply a juvenile ape. Sts 5 demonstrated without doubt that had the Taung Child grown up, it would not have developed into an ape.

Thus Dart and Broom's long-suffering campaign for recognition paid off 22 years after the Taung skull controversy had first surfaced. Their rehabilitation, which began with the Americans and then Le Gros Clark, was finally confirmed in the form of a resolution taken by the British Association in 1947. It offered Broom "its congratulations on the brilliant success of his recent exploration of the Sterkfontein site. His new discoveries amplify and confirm in a remarkable way his interpretation of the earlier finds and also provide a vindication of the general view put forward by Professor Raymond Dart in his report of the first *Australopithecus* skull found in 1924."

Savoring this victory, Broom switched his attention in 1947 to Swartkrans. He managed to raise funding from a wealthy young American named Wendell Phillips, who led and financed the California African Expedition in a search for hominids in South Africa. This expedition visited every hominid site in South Africa during 1946 and 1947 but found no new fossils. Broom dismissively said of Phillips that he "had more money and egotism than scientific ability," but nonetheless accepted the contributions. The Americans did not have much joy, but later that year at Swartkrans, Broom found an adolescent hominid mandible containing five cheek teeth and three molars (SK 6). He noted that the massive jaw and thickly enameled teeth were different from the previous Sterkfontein finds and, as Donald Johanson says, Broom "succumbed to his usual temptation to give an entirely new genus name to the jaw." He called it *Paranthropus crassidens* (from the Latin *crassus* meaning "thick" or "solid" and *dens* meaning "tooth"), but it was eventually classified as *A. robustus*.

At the time, Swartkrans was being mined for calcite, so Broom was forced to inspect the piles of cast-off breccia for specimens. On June 30, 1950, he found in such a pile a "fine skull" (SK 48) that had been blasted out of the rock. Although it was damaged in the explosion, the jaw was largely intact and helped shed more light on the distinctive dental morphology of the robusts and, until recently, was the most complete early hominid skull from South Africa.

The confusing jumble of names given to the early australopithecines was partially cleared up by Broom, who spent his last years trying to make sense of the ape-men species. He concluded after a lengthy study that there were two main types of australopithecines: the robust and the gracile, both of whom had walked upright. And although they had relatively large ape-like faces, their teeth were humanlike. Broom believed the australopithecines could use rudimentary tools, a source of debate even today. But the robusts seemed to be more primitive than the graciles, even though they were not as old. Did they represent, then, an evolutionary cul-de-sac? Tobias has written: "Since the two ape-men seemed clearly to be on the line of human descent and the other to have specialized away from that line, Broom's finds compelled scholars to realize that not all early hominids were direct ancestors of modern mankind. Some were on side branches. This meant that at an earlier period the two species, so closely related to each other, must have branched off from a common ancestor. The pattern of hominid evolution was not like a linear Chain of Being after all. It was like a bush of branches, only one of which made the grade to the later stages of human evolution, while the other branches were doomed to ultimate extinction."

Broom died in 1951, driven fanatically to the end to try and complete his monograph on the Swartkrans hominids. He completed the final corrections on April 6, 1951, and reportedly said: "Now that's finished…and so am I." He passed away that evening.

CHAPTER 4

EAST SIDE STORY

~

The wheels of life often spin on rims of irony. Just as Robert Broom was gaining international recognition for the South African quest for human origins, circumstances within the country conspired against him. South Africa has always been—and still is—an intensely turbulent political society. In retrospect it is quite remarkable how, in just a few years, the post-World War II dynamic unraveled half a century of scientific endeavor. Broom spent his last years consolidating South Africa's contribution to the understanding of our ancestry, which in simple terms can be summed up in two broad themes: first, that Africa was humankind's original home, and second, that at least some of the australopithecines were an intermediate species between ape and man. By 1951, when Broom's long life came to an end, South Africa was still the only country where fossil remains of australopithecines had been uncovered—and all five sites containing the fossil evidence were in dolomitic limestone caves.

Broom should have been regarded as a national hero. After decades of debate, the world's scientists had finally accepted the importance of Taung and recognized Broom's accomplishments in unearthing not only an adult specimen of *africanus* but at least one other species of ape-man. Instead of being honored, however, Broom became a victim of the politics of his time. The support that he enjoyed from the highest level of government disappeared overnight as Prime Minister Jan Christian Smuts was toppled from power. General Smuts's United Party government, which had taken South Africa into World War II on the side of the Allies, suffered a backlash at the hands of Afrikaner nationalism. There had been debate within the white community on whether or not South Africa should have entered the war on the side of the British. Many Afrikaners regarded the British as mortal enemies from the time of the Boer War, which had pitted Queen Victoria's empire against Afrikaans nationalists at the turn of the century. Smuts had fought on the side of the Boers but subsequently accepted British Commonwealth status for South Africa. His decision to go into the War had fomented his political opponents. A resurgence of Afrikaner nationalism saw Smuts swept from office in the 1948 elections. With him went the proactive official support for the exploration of human origins that was enjoyed since the discovery of Taung.

The theory of human evolution was a direct philosophical challenge to this new government, which over time crystallized its policies into what we today know as apartheid—strictly enforced racial segregation at all layers of society. Until 1994, all black South Africans were denied the right to vote. The National Party is unashamedly creationist in its views, as reflected in Christian National Education school policy, which barred the teaching of evolution or any other kind of science that suggested there was no biological justification for racism. To suggest that whites and blacks may have evolved from the same common ancestor, and that this ancestor was an ancient African ape, was an effrontery to the white supremacist attitude of South Africa's new rulers. The Second African Prehistory Conference, which was scheduled to be held in South Africa in 1951, was relocated to Algeria because the National Party gave notice that it would not allow black delegates to attend. This was a portent of things to come.

The implementation during the fifties and sixties of what became known as Grand Apartheid led to South Africa's increasing political isolation from the rest of the world and the ensuing academic boycott, which resulted in drying up research funds. Although the search for human origins continued under universities like Wits and UCT, the new government frowned upon and gave little official support to the digs at Sterkfontein, Kromdraai, Makapansgat, and Swartkrans.

It is hardly surprising, then, that East Africa became the new locus for the exploration of human origins during the 1950s. Contrary to the prevailing political mood in South Africa, the new rulers of the independent East African states seized on the fact that their region held fantastically rich evidence of human prehistory. As apartheid South Africa became more profoundly set off from the world and neighboring countries on the continent, the importance of Kenya, Ethiopia, and Tanzania became apparent in proving to the rest of the world that Africa was indeed the cradle of humanity.

Geologically, the cradle was the Great Rift Valley. This 1,860-mile-long gash in the landscape is the remnant of a monumental earthquake that tore Africa 20 million years ago. The Rift stretches like a great scar up the eastern half of Africa, snaking northward from Mozambique through Malawi, Zambia, Tanzania, Burundi, Rwanda, Uganda, Kenya, and into Ethiopia and northern Somalia. Satellite photographs show how neatly it divides the coastal plains and towering highlands of the eastern seaboard from the rest of the continent. The Rift embraces some of Africa's most spectacular scenery, juxtaposing great ancient volcanoes with their snowy caps against the rolling game-rich savanna of the lowlands and the broken woodland of the escarpment.

For the first few million years of its existence, the floor of the Rift was probably under water. Then, five million years ago, as the Pliocene ice age set in, sucking up much of the moisture that had characterized the Miocene, this great inland sea dried up, leaving behind the string of great lakes that are Africa's geographical landmarks. These are the remnants of what were once giant inland watercourses. Fossils in the Rift have been well preserved, generally sandwiched between volcanic sediment that lends itself to accurate dating. This has provided an invaluable timeline to the scientists who prized the australopithecine and *Homo* fossils extracted from these sediments during the sixties, seventies, eighties, and early nineties.

The Rift is thus an archaeological hotline to Africa's past. The ecology of this valley system may have created the conditions that favored the emergence of our earliest human ancestors. That, in any event, was the view of Professor Yvves Coppens, the French archaeologist who dubbed this theory the East Side Story. His reasoning went along the following lines: The jagged hills created by the Rift provided a natural barrier for the moist winds blowing inland from the Indian Ocean. This created a high rainfall belt along the foot of the eastern highlands, which became a buffer against the expanding savanna. The great Miocene forests, which had once covered most of the African landmass, retreated to the tropical confines of the equatorial belt. These African tropical forests remain today and are the last surviving habitat of the gorillas and chimpanzees that were once part of the hominid ancestral lineage.

Coppens argued that the primate forebears of humans and apes were geographically split by the Rift: "The western descendants of these common ancestors pursued their adaptation to a life in the humid arboreal milieu, these are the [apes]. The eastern descendants of these same common ancestors, in contrast, invented a completely new repertoire in order to adapt their life in an open environment: these are [humans]."

Yvves Coppens, a genial man now based at the College de France in Paris, admits that his theory has largely been invalidated by other developments since the sixties, when it was originally formulated. For a start, the remains of australopithecines, originally thought to have been confined to eastern and southern Africa, have now been found in West and North Africa. Coppens, who has been intimately involved in the description of West African australopithecine fossils found in Chad in 1995 by Michel Brunet, believes that these relatively recent discoveries indicate that there were regional variations of ape-men that lived across the continent. Theoretically, any one of them could have given rise to the genus *Homo*. But for the past half century, the conventional wisdom has been that East Africa had been the cradle of humanity. This may still be so. There is no disputing that the oldest hominid fossils are from East Africa, and that despite the complications that have clouded the search for our original ancestor, the Rift contains vital pieces of the paleontological puzzle. So even though the East Side Story theory of evolution is outdated, it is an apt enough title for the drama of the East African hunt for the earliest hominids.

The prehistorical importance of the Rift first became apparent with the European discovery of Olduvai Gorge, an arid branch of the valley system that cuts through northern Tanzania to the Kenyan border. In 1911, a German entomologist named Kattwinkel literally stumbled upon Olduvai as he chased butterflies across the Serengeti plains. Climbing down the sides of the valley that he had almost fallen into, Kattwinkel noted the number of different exposed sediments and was struck by the numerous fossil bones clearly visible on the valley floor. He provided the first written record of East African fossils.

Then in 1927, while the academic world was embroiled in the Taung controversy, a settler in western Kenya found some fossils embedded in a limestone quarry and sent them off to the British Museum. There they were studied by Tindall Hopwood, who recognized them as a "prehuman" tooth and jawbone. Hopwood organized an expedition to East Africa in 1931 and recovered two more fossils from the same area. Two years later he announced that he had found the "chimpanzee's ancestor," a species he called *Proconsul africanus*. In doing so, he displayed a sense of humor rare amongst scientists, because Consul—his name, from *Proconsul*—was a chimp at the London Zoo that delighted spectators during the 1930s with his ability to perform tricks such as riding a bicycle and smoking a pipe.

At about the same time in Europe, there was renewed interest in a modern-looking hominid originally found by German paleontologist Hans Reck in 1913 in ancient sediments at Olduvai Gorge. This attracted the attention of a man whose name has become synonymous with the history of East African paleontology, Louis Seymour Bazzet Leakey. Born to English missionary parents in Kenya in 1903, Leakey grew up among the Kikuya and was, at the age of 13, initiated into the tribe. As a white man with a deep empathy for black Africa's beliefs and customs, he carved a unique international reputation for himself as a social anthropologist, paleontologist, naturalist, and conservationist. With his second wife, Mary, he pioneered the search for human origins in East Africa, a legacy taken up by his son Richard and daughter-in-law Meave.

Louis Leakey did not believe that the australopithecines were humankind's immediate ancestors. He, like Sir Arthur Keith, believed in an ancient occurrence of "true man," and that the ape-men and even *Homo erectus* had diverged from the human stem and had eventually "perished without issue." Leakey's stubborn insistence in holding on to this view throughout most of his life set him on a collision course with many other scientists—among them, his wife, Mary. It is understandable, then, that Leakey was excited by Reck's find, which at first seemed to prove his theory. On closer inspection of the site, however, it was discovered that the skeleton was actually that of a modern Masai warrior who had been buried in very old sediments. He overcame this early setback and with Mary led a number of expeditions to Olduvai and to sites around Lake Victoria. They amassed a number of interesting primate fossils—including the first almost complete *Proconsul* skeleton in 1948—and countless stone tools.

But it was only in 1959 that the husband and wife team finally struck paleontological gold at Olduvai. On July 17, 1959, almost 100 years exactly after Darwin published *On the Origin of the Species*, Mary Leakey discovered a new hominid species. Louis had remained at the camp with the flu, while Mary went out with her two Dalmatians to examine a site where a number of mammal fossils had been found on the surface, recently exposed by heavy rains. As she later wrote:

> But one scrap of bone that caught and held my eye was not lying loose on the surface but projecting from beneath. It seemed to be part of a skull including a mastoid process (the bony projection below the ear). It had a hominid look, but the bones seemed enormously thick—too thick, surely. I carefully brushed away a little of the deposit, and then I could see parts of two large teeth in place in the upper jaw. They were hominid.

She rushed back to call Louis to see the specimen. He was initially disappointed it was not *Homo* but soon became aware that it was a remarkable find, the first australopithecine found outside South Africa. Leakey later named the specimen *Zinjanthropus boisei*, deriving its name *Zinj* from the old Arabic name for East Africa and *boisei* after the financier of the Olduvai work, Charles Boise.

The discovery of the specimen was announced at the Third Pan African Congress, which was held in Leopoldville (now Kinshasa) in the Congo in August 1959. Louis and Mary had flown there from Nairobi with the skull in a box on her knee. At the Congress, Zinj—also known as "Dear Boy" to his immediate circle of admirers—attracted enormous attention, although many delegates took issue with Louis for creating a new genus: The skull seemed similar to the robust australopithecines that had already been found in South Africa. There was also some discussion as to whether Zinj had been a toolmaker because of the number of stone artifacts found in association with the skull at Olduvai. By this time, advances in dating techniques, particularly the use of potassium–argon decay, enabled greater accuracy in fixing a date on fossils, and Zinj was estimated to be 1.8 million years old.

Louis Leakey offered Phillip Tobias the opportunity to prepare a technical report on the skull, and Tobias promptly placed Zinj firmly back in the australopithecine camp, although he conceded that it was a new species. Zinj was thus reclassified as *Australopithecus boisei*, but was commonly known as the Nutcracker Man after a comment at the congress made by Tobias, who, on glancing at its dental structure, said in awe that "he'd never seen such a set of nutcrackers." The Nutcracker Man enabled the Leakeys to raise considerable finances from the National Geographic Society, which in 1960 contributed $20,200 for further field excavations in exchange for the exclusive American publishing rights for NATIONAL GEOGRAPHIC.

Barely a year later came another significant discovery from Olduvai. Leakey had assembled a team, including Tobias and British paleontologist John Napier, to examine a number of other fossils that had been recovered from Olduvai. In particular four partial skulls were identified, and on the whole they seemed to have far greater cranial capacity than any of the South African australopithecines found thus far. They also had more humanlike teeth and a less protruding face, while some of the associated hand bones showed that these creatures were capable of a precision grip, suggesting that they were toolmakers.

After three years of analysis and discussion, the description of the find was published in *Nature* in 1964, with Tobias, Napier, and Leakey claiming that they had found a new species within the genus *Homo,* which they called *Homo habilis*— "man that is able." The name *habilis* was suggested by Dart to denote its tool-making capacity, and it was soon dubbed in the press as the "Handy Man." *Habilis,* claimed the authors, was intermediate between the australopithecines and *Homo erectus.* They concluded that the specimens were 1.8 million years old. This made them roughly contemporaneous with Zinj and his ilk. Leakey for one was more comfortable claiming that *habilis* was the original toolmaker and not Zinj.

Habilis's reception was predictably stormy. The fragmentary nature of the fossils troubled many scientists, who doubted that there was sufficient evidence to justify a new species altogether, never mind the bold claim that this was the first member of the *Homo* family. Additionally, there was a problem of classification around the so-called cerebral Rubicon. At what stage does an ape-man become human? The conventional wisdom of the time had it that *Homo* had a cranial capacity no less than 700 cubic centimeters, an arbitrary enough measurement based on an examination of *erectus* skulls that had been found. The average for the *habilis* specimens was 642 cubic centimeters according to Tobias. This was still about 200 cubic centimeters greater than the average australopithecine, which in his estimation justified a new species. Others, including his South African colleague J. T. Robinson, believed that *habilis* was a regional variation of the gracile australopithecine. Robinson and Le Gros Clark argued that there was not enough room between the australopithecines and *erectus* to warrant a new species, particularly given the lack of postcranial remains. Although *habilis* has been scientifically accepted as a distinct species, it sits rather enigmatically in an uncomfortable taxonomic niche. During the 1980s, the *habilis* debate reared once again over two finds—the large-boned KNM-ER 1470 and the tiny OH 62, both of which were classified as *habilis* but which display very different characteristics.

In 1966, Emperor Hailie Selassie of Ethiopia met Louis Leakey during a state visit to Kenya and viewed some of the Olduvai fossils. When asked why there weren't any such artifacts in his own country, Selassie was told that there undoubtedly were, but that there were problems in getting official permission to undertake such a search. Leakey's response to the Emperor wasn't a wild guess. He had collaborated with American paleontologist F. Clarke Howell's reconnaissance trip to the Omo region of Ethiopia a few years before, and was aware that a French expedition to the region in the 1930s had reported finding a veritable treasure chest of fossils. The Emperor soon sent an official invitation to Leakey, who decided to assemble an international expedition involving Kenya, America, and France. Leakey was too old to participate personally, so he nominated his 23-year-old son, Richard, to represent him, while F. Clarke Howell headed up the Americans and Yvves Coppens led the French team.

The planning of the Omo Research Expedition, which worked in Ethiopia from 1967 to 1974, was significant in that it represented a new approach to fossil hunting. This was a departure from the days in which "loners" like Louis Leakey scoured the landscape, finding fossils and afterward calling for expert help in making sense of it all. The Omo expedition consisted of a team of individual specialists in a variety of disciplines; each find would be analyzed from a variety of scientific perspectives. The expedition was also significant in that Omo turned out to be a geological yardstick for the last four million years. This was because of the volcanic ash fall that created layers in which plant, insect, and mammal fossils were found, meaning that groups of fossils could be dated together. The Omo expedition found four kinds of hominids during the seven field seasons it spent in Ethiopia, the most abundant being the massive robust australopithecine closely resembling Zinj, traces of a gracile ape man, *habilis,* and *erectus.* It was forced to close shop when Hailie Selasie was overthrown by Marxist guerrillas.

Richard Leakey, however, had abandoned the Omo expedition fairly soon after it started. Sensing that he was a junior partner in the proceedings and that he would get little credit for any discoveries, he borrowed a helicopter and flew over southern Ethiopia and northern Kenya to look for his own fossil site. This turned out to be the area around Lake Rudolph, now known as Lake Turkana, which proved to be as rich as Omo in hominid fossil material. Leakey set up his headquarters at Koobi Fora with the backing of the Kenyan government and NATIONAL GEOGRAPHIC. Louis was shocked at his son's decision to break away from the Omo expedition but was pacified eventually by the discovery of plentiful hominid fossils at Koobi Fora.

Richard Leakey is a man who has survived personal extinction—despite a near-fatal light aircraft crash, a controversial tenure as a leading conservationist in which he led the campaign to ban ivory trading, and a foray into the world of opposition politics in Africa. His critics have often accused him of arrogance and stubbornness, borne from his family's reputation of paleontological aristocracy. There is no doubt that Richard exhibits a fierce individualism, a streak that was beginning to make itself felt even as a 23-year-old. Having initially shunned paleontology as a career so as not to be trapped in the shadows of his famous parents, he then took the science up with a vengeance.

These traits of strong individualism saw the cohesiveness of the Leakey family begin to shred during the late sixties. Besides Richard's fallout with his parents, Mary and Louis themselves entered a period of separation. One of the main differences they had was political. Mary's decision to accept an honorary degree from Wits in 1968 was criticized by Louis, who refused to attend the ceremony because of South Africa's apartheid policies. Later, Richard was to cross swords with his brother Phillip, who entered politics as part of the ruling party. Their relationship improved somewhat after Phillip generously donated one of his kidneys to Richard in the mid-1970s.

In August 1972, three months before his father died, Richard Leakey found one of the most significant fossils of the century, KNM-ER 1470, a well-preserved *Homo* specimen that was originally dated to three million years. Although subsequent analysis has placed it about a million years younger, 1470 represented a far older form of *Homo* than had been found previously. It was also older than most of the australopithecine finds, which challenged the thinking that the ape-men were ancestral to *Homo*. Although Leakey himself was reluctant to assign it to any species, instead putting it into what he called a "suspense account," it soon became classified as *Homo habilis*. However, 1470's brain capacity of 800 cubic centimeters raised other questions about *habilis*, and again fueled the debate as to whether Louis Leakey's 1960 specimen, which had a brain capacity almost 200 cubic centimeters smaller, could be classified as the same species. Whatever the classification, the significant feature of 1470 was that it crossed a line between ape and human. Phillip Tobias believes that its brain size and morphology enabled it to have the neurological basis of speech. This, according to Tobias, may not have been a recognizable spoken language, familiar to modern humans, but a mixture of noise and sign language.

In 1973, a new area of Ethiopia opened up to science. This was the Afar Triangle, a hostile and barren landscape in the Hadar region, straddling the Great Rift Valley as well as two other minor rifting systems. Johanson, who had participated in the 1970 and 1971 field seasons at Omo, visited the area with the Frenchman Maurice Taieb and was convinced that this was good fossil territory: "It was a place that paleoanthropologists only see in their dreams," he wrote. The International Afar Research Expedition was established with U.S. funding, and Johanson's team was joined by Taeib and Coppens. During that first field season, they found a three-million-year-old hominid knee joint, which was the oldest evidence of bipedalism so far. Johanson was starting to make a name for himself.

The following field season in the Afar yielded Lucy. The story of the find has now become well known. Johanson woke up on the morning of November 30, 1974, and wrote in his camp diary that he was feeling lucky. He and his colleague Tom Gray had spent several hours searching a presurveyed area near their camp when they decided to look in a gully before returning to shelter to avoid the blistering midday heat. Johanson noticed a piece of primate arm bone on the surface of the ground. As the two men kneeled over the bone, arguing whether or not it was a hominid, they noticed that the bones of an almost complete skeleton surrounded them. In a state of high excitement, they realized that it was a hominid, a very old one at that, and raced back to camp bursting with their news. The entire team returned in the afternoon and picked up several hundred pieces of bone representing about 40 percent of an individual skeleton. As Johanson wrote in his bestseller, *Lucy*:

> But a single individual of what? On preliminary examination it was very hard to say, for nothing quite like it had ever been discovered. The camp was rocking with excitement. That first night we never went to bed at all. We talked and talked. We drank beer after beer. There was a tape recorder in the camp, and a tape of the Beatles song "Lucy in the Sky with Diamonds" went belting out into the night sky, and was played at full volume over and over again out of sheer exuberance. At some point during that unforgettable evening—I no longer remember exactly when—the new fossil picked up the name of Lucy, and has been so known ever since.

More was to come. The following year Johanson's team, accompanied by a NATIONAL GEOGRAPHIC photographer and a French movie team, trawled the Afar area looking for more hominid fossils. A member of the film crew, a woman named Michele, who Johanson described as enthusiastic but incapable of telling a "fossil from a Coca-Cola bottle," sat down in the shade of an acacia bush and nearly sat on top of the remnants of what became known as the First Family. This was a collection of 200 bones from no less than 13 individuals on a bone-strewn slope.

The Hadar collection, as these fossils are known, were shipped back to California for analysis. Enlisting the services of Tim White, the bright young American paleontologist who'd worked with the Leakeys, Johanson began an exhaustive process of determining what he had found. Lucy's skeleton consisted of 47 of the 207 bones in the body including parts of the upper and lower limbs, backbone, ribs, and pelvis. The fragments of her skull that were found indicated that her brain capacity was small, measuring probably around 420 cubic centimeters. Lucy's skull was not as complete as Johanson would have liked, and the discovery of the First Family (collectively known as AL 333) allowed a closer look at the cranial morphology, facial shape, and brain capacity.

Lucy's face would have been prognathic, jutting out almost to the same degree as a modern chimpanzee, while her teeth were a mixture of primitive and derived traits. She was a mature adult, standing just over three feet tall, and she walked on two legs much like a human. Well sort of. This has been a point of contention, with Johanson's colleagues divided as to whether Lucy was fully bipedal or whether she was predominantly tree-living and unconfident of her two-legged gait. Of one thing Johanson was certain: Lucy represented the oldest hominid yet discovered, dating back 3.1 million years. Johanson and White concluded that Lucy was close enough to the South African australopithecines to be included in that genus, but represented a new species altogether. She was christened *Australopithecus afarensis* in honor of the geographical location in which she was found.

One of the challenges in analyzing the First Family was the variation in size of some of the individual specimens. This was ascribed by Johanson and White to sexual dimorphism: the difference in size between males and females. Paleontologists like Richard Leakey were dubious. Leakey said he thought that the Hadar collection might represent more than one kind of ancient hominid, and he stuck his neck out enough to say that the bigger specimens in the collection were an ancient form of *Homo*.

Johanson conceded the problems of classification: "It was difficult at first to find a place for Lucy on the hominid family tree, but few today would deny her an important role in hominid evolution. To some she is the 'mother of all humankind' and to others she is the 'woman who shook up man's family tree.'"

Whatever Lucy's nomenclature, it had become apparent that bipedalism has been around for well over three million years. It soon became apparent that this estimate could be pushed back to almost four million years after Mary Leakey's expedition at Laetoli in Tanzania uncovered an astounding set of footprints preserved through a fluke of nature.

Mary Leakey had begun excavations at Laetoli in 1974, but it was only two years later that the real value of the site became apparent. Late one afternoon, some of the younger members of her party were fooling around, throwing elephant dung at each other in a dry gully near the camp. One of them, Kenyan Museum paleontologist Andrew Hill, ducked a dung missile and, as he stooped to look for something to throw back, noticed some unusual dents in the layers of hardened ash that formed the bed of the dry stream. On closer inspection, these dents looked like mammal footprints. It was only in 1977 that serious survey work began in the area and the prints were indeed found to be those of a variety of animals, including elephants, antelope, giraffe, rhino, pigs, baboons, and a number of birds.

What seemed to have occurred was that a nearby volcano, Sadiman, had spewed clouds of carbonate ash during an eruption 3.7 million years ago. This volcanic powder had settled over the surrounding landscape and was soaked by rain, giving it a cementlike consistency. The baking sun hardened these impressions, which were then given further protection by another series of mild volcanic eruptions, effectively creating a multimillion-year natural sealant. In time, these prints were covered up by soil but did not lose their original shape. Coincidentally, they became exposed again through natural erosion during the Laetoli excavations. One set of prints looked suspiciously like they might be human. Mary Leakey selected one of her Kenyan staff members, Ndibo Mbuika, to further excavate around the prints, and he soon called her over to show her what he'd found. John Reader, in his book *Missing Links*, sums up that moment on August 2, 1978, at 10:45: "Mary Leakey straightened up abruptly. She lit a cigar, leant forward again, scrutinized the excavation before her and announced, 'Now this is really something to put on the mantelpiece.'"

Mary Leakey entrusted Tim White to continue excavations and to begin taking casts of the prints that had been uncovered. What emerged was a series of about 50 footprints over a distance of 72 feet, caused by two or three hominids who'd walked upright across that sandy riverbed. Although there has been some controversy as to whether the prints were made by two or three hominids, what is remarkable is that their shape and gait suggested by the distribution of weight and indentations of the toes, ball, and heel, are surprisingly close to that of modern humans. White believed that they were made by the same species as Lucy. What Laetoli has given us is proof that bipedalism arose in Africa at least four million years ago.

There was another controversy around Laetoli, particularly in the way the prints were to be preserved. They were covered with soil to protect them from the increased number of visitors who were coming to see these ancient footsteps. Unfortunately, the soil was insufficiently sterilized and some of the prints were destroyed by seeds that rooted. The site was reburied with funds from the Getty Conservation Institute in 1996, using sand layered with geo-textile materials to prevent additional damage. They will not be uncovered again for another 50 years. The scientific community remains concerned, however, about the Tanzanian government's ability to protect the site and to halt further degradation of this remarkable find.

The string of discoveries of the sixties and seventies was beginning to suggest a fairly straightforward human family tree topped by the Laetoli-Hadar hominids (*afarensis*) branching into *africanus* and *Homo*. But during the eighties, three finds in particular began complicating this scenario.

Richard Leakey's team of fossil finders discovered the remains of an athletic-looking hominid boy on the banks of Lake Turkana in northwestern Kenya. Kamoya Kimeu, a stockily built Kenyan who was a leading member of Leakey's Hominid Gang, came across a small piece of hominid skull on a lava pebble–covered slope near the Nariokotome camp. "Kamoya's skill at finding fossils is legendary," wrote Leakey. "A fossil hunter needs sharp eyes and a keen search image, a mental template that subconsciously evaluates everything he sees in his search for telltale clues. A kind of mental radar that works even if he isn't concentrating hard."

The Turkana Boy, as Kamoya's find became known, was extraordinary on two counts. First, the rest of that field season yielded a virtually complete skeleton, and second, he defied the preconception of the time in terms of size. Traditionally it was thought that hominids started off short and squat—Lucy after all would have been just over three feet tall—and then through the course of human evolution, they would have grown taller and bigger-boned. The Turkana Boy, however, negated this. Despite his antiquity of 1.6 million years, he had body proportions that would be recognizable in modern humans. Although classified *erectus*, he was notably different from the squatter examples of that species that had so far come to light. Had he lived to adulthood, the Turkana Boy would have grown into a strapping six foot man. What was puzzling is that later hominids did not match up to these proportions. Was the Turkana Boy a freak, or was there some other as-yet unconsidered factor in evolution being played out?

The following year another find that challenged preconceptions emerged from the Leakey camp. This was the Black Skull. It was discovered by Richard's close colleague, English paleoanthropologist Alan Walker, near Nariokotome while the Hominid Gang was reconstructing an ancient hippo skull. Walker saw a dark-colored fossil next to a pile of stones, and noticing that it was part of an upper jaw with enormous tooth roots, his first thought was that it must be a bovid. "Then I saw another piece," Walker wrote, "from the front of the skull and thought it was a big monkey. But then I turned it over and saw a frontal sinus—hominid."

Walker then casually informed the rest of the team, including veteran Hominid Gang–member Kamoya. "If I show you a three-million-year-old hominid will you get me a beer?" Walker asked Kamoya back at the camp. Kamoya thought he was joking and went off to get the beers anyway. When he returned, Walker gave him the collection bag and watched him open it. "Look at the size of those tooth roots" was all that Kamoya could say.

"The new skull is going to force people to change their mind," Leakey recorded in his diary. "I am even more convinced that the three hominids of two million years ago are going to be traced back to three million years. Also it is likely that Johanson-White can be shown to be very wrong in their scheme. We shall see. It's going to be very interesting for some."

What gave rise to Leakey's smugness was this: Johanson and White believed that one of the descendants of *afarensis*, dating back around 3.4 million years ago, was *africanus*, which at 2.5 million years ago gave rise to the robust australop-ithecines. But the Black Skull was dated to 2.5 million years ago, making it contemporaneous with *africanus*. Where then did the Black Skull come from? Could *afarensis* have given rise to the robust lineage of ape-men? As Leakey noted, the skull "would make the human bush even bushier, our evolutionary pattern more complicated."

There was predictable debate over the classification of the Black Skull (KNM WT 17000). With a primitive brain case of 410 cubic centimeters, it was one of the smallest of all known hominids. Eventually it was classified as *Australopithecus aethiopicus*, a new species of robust ape-man, and was believed to pre-date *A. boisei*. Certainly the Black Skull was paleontologically entertaining, as my colleague Henry McHenry observed: "No one could have predicted this combination of features. It turns a lot of our ideas upside down, about the sequence of evolutionary changes in the skull and about who is related to whom.

The most significant concluding find of the late eighties came from the Johanson-White camp in the form of OH 62. Discovered by Tim White in July 1986 close to the heavily trafficked tourist route to Olduvai, OH 62 was found in a 48-square-yard area filled with 18,000 various mammal bone fragments. It is testimony to Tim White's patience that he sifted through these to identify 302 bones belonging to OH 62, which seemed to bear the features of *Homo habilis*, particularly in the dentition. The lack of an even partially complete skull complicated the analysis, however, and the fact that the creature would have been almost ape-manlike in stature, made many question its classification in the genus *Homo*. As Leakey himself said rather disparagingly: "Johanson's got himself more into an evolutionary hole than an intellectual knot," claiming that either 1470 or OH 62 could be classified *habilis*, but they had too many differences to belong to the same family. What OH 62 did have, however, was a good spread of upper- and lower-limb parts. From my perspective, what was so significant about this little 1.8 million-year-old creature was that it apparently had long arms and short legs!

By the end of the eighties, Leakey had all but abandoned paleontology, taking up a position as the head of Kenya's wildlife services. With Kenyan President Daniel Arap Moi, he made headlines in 1989 by stage-managing the burning of ivory worth three million pounds sterling that had been confiscated from poachers to kick-start a successful campaign to ban ivory trading. The two later had a falling-out when Leakey challenged the authorities by blocking the construction of an oil pipeline through Nairobi's game park, a move that eventually forced him to resign in March 1994. That same year, his Cessna 206 inexplicably stalled on takeoff from Nairobi's Wilson Airport during a routine wildlife inspection. Leakey was lucky to survive the crash, which later prompted unproven allegations of sabotage. Badly concussed and barely conscious, he was taken to a mission hospital in a farmer's truck.

Leakey was transferred to Nairobi Hospital, where he remained for 12 days, hovering close to death. As his condition deteriorated, his wife, Meave, in desperation contacted their friend and supporter Queen Beatrix of the Netherlands. She acted promptly by paying for orthopedic trauma specialist Professor Christopher Colton to fly to Nairobi and help save him, which he did through the course of 14 operations. Richard, however, lost both his legs, and today he walks on prostheses. He said in a magazine interview in 1995 that his real legs had been embalmed and stored in a suitcase at his home in preparation for a proper burial at Koobi Fora, which is where he has asked to eventually be laid to rest.

Mary Leakey died in December 1996, aged 83. Paying tribute to his mother, Richard Leakey said: "In the 1950s and 1960s Louis [his father] got most of the publicity, probably because of the chauvinism of the time. But Mary was the centerpiece of the research." This point was echoed by E. Barton Worthington, one of the first scientists to explore Lake Turkana in the 1920s, who said that "Louis was always a better publicist than scientist. Mary was the real fossil hunter."

CHAPTER 5

FOSSIL FORENSICS

~

The search for fossils in some ways is similar to police detective work. There's the "scene of the crime"—the site where a death has occurred—which has to be cordoned off from bystanders so investigators can search unhindered for clues. The context in which the body has been found has to be carefully examined before it is removed for further study. The integrity of any evidence found at the scene has to be protected and a detailed layout of the location prepared. The services of a specialist may even be required in order to ascertain the motive for the killing. Without taking these basic forensic steps, the investigator does not have a case. So, too, with the evolutionary detectives.

The common skill required by both paleontologists and police is that of forensic observation. Indeed, Louis Leakey's acute powers in this regard were used extensively by the Kenyan police during the 1940s. He assisted the Central Investigation Department in Nairobi in a number of cases, ranging from the suspicious circumstances around the death of the flamboyant Lord Errol (the subject of the book and film *White Mischief*) to the trial of a missionary charged with murdering his mother-in-law. Leakey's experience with casting fossils helped solve a murder case in which an army truck driver was accused of killing a young woman.

Leakey made casts of the footprints found at the scene of the murder and followed them into a nearby swamp. The casts enabled the police to trace the man, while one of his colleagues at the Kenya National Museum, the botanist Peter Bally, identified a rare plant seed from the swamp that had stuck to the accused's trousers while he was making his getaway.

When most people think of fossil human ancestors, they probably think of perfect skulls with gleaming teeth lined up on a museum shelf. In fact, a complete skull of an early hominid is almost a myth. This is because violent death is a messy business that does not respect cosmetic niceties. Most of the old hominid material we have today entered the fossil record through unpleasant means. The great majority of primate skeletons we find both in eastern and southern Africa are from individuals who suffered a brutal death, usually through the violent actions of some predator. Their remains are usually scavenged and then, once the feasting is finally done, what is left must undergo the punishment of the fossilization process. A further complication is the relative fragility of human—and primate—bones compared to those of other mammals.

There are many factors that mitigate against the preservation of fossil material. But even when we find the evidence of ancient bones, it's rarely in a manner that allows for easy assemblage of the original form. Usually there are splintered fragments that have to be painstakingly pieced together, and even if this is being done by the most skilled paleontologist, there is still a considerable margin of error. This is particularly true of the hominid skull, which is fragile at the best of times and consists not of just one bone but of more than 28. A slight difference in the angle of reconstructing a fossil skull can have huge implications given how important cranial capacity is in trying to determine how smart, or how human, various hominids were in the distant past.

Richard Leakey's colleague Alan Walker found that the skull of the Turkana Boy had been exposed to the surface through natural erosion. But then this same process had allowed the seed of an acacia tree to be deposited into the upturned, soil-filled cranium, a perfect potting structure if ever there was one. More damage was done to the skull by the 20 years of growth of that acacia tree than in probably a million years of secure burial in ancient sediments. It took Walker and Leakey the better part of a field season to find all the bits of skull that had been fragmented in a slow-motion dispersal by the tree's growing root system. Sometimes the damage to bones is even more immediate. Mary Leakey told the story of one of the original *Homo habilis* skulls she found at Olduvai in the early sixties. Her team had discovered the skull fragments late in the afternoon, after they'd been exposed in a riverbed by heavy rains, and decided to begin work on them the next day. During the night, a herd of Masai cattle trampled right over the bones, splintering them to smithereens.

The recovery of any complete bone of a hominid fossil is so rare that it is considered an exception. This is why there is so often controversy around discoveries that involve the naming of new species. A single four-million-year-old fossilized hominid mandible from Lothagam in Kenya is the only trace of one of our earliest ancestors, but who would be brave or foolish enough to try and name a genus or species on this paltry evidence. There are of course exceptions. Much is made of how fantastically preserved the Lucy skeleton was, but then she is only about 40 percent complete. Even more of the Turkana Boy skeleton was eventually recovered, but a critical number of bits and pieces are still missing from this *erectus* juvenile's fossilized frame. There have been a handful of whole limb bones found, complete from one end to another, but most have been discovered in isolation from the rest of the body and are typically very rugged bones, such as the femur.

The hardest and therefore most durable part of the hominid body is enamel, which means that teeth are the most likely part of the body to be preserved in good condition. It is not surprising then that so much of the fossil collection—thousands of specimens—consists of teeth. In fact, about 80 percent of the entire South African fossil record is made up of isolated teeth, and the same is probably true for East Africa. One can see how dependent scientists have become on dental analysis for charting the body's gradual change from ape to human ancestor. Teeth are therefore a very handy tool for measuring evolution. Giving us information beyond just genus and species, they tell us about an individual's age, health, and diet: Raymond Dart could guess the age of the Taung Child at its death because the milk teeth were still erupting; microscopic analysis of robust australopithecine teeth tells us their diet consisted of rough vegetation; Lucy's wisdom tooth was starting to wear, which indicated that, despite her tiny frame, she was a fully grown adult. The study of teeth—the body's main food-processing mechanism—is a whole other area of specialization in paleontology.

Clearly, a knowledge of anatomy is a prerequisite to evolutionary detectivehood. But the nature of paleoanthropology is such that other forensic skills are also needed to piece together the clues of an ancient "crime." It is easier to speculate on how a gangster ends up in the Hudson River with concrete shoes than on a body that has been lying in a cave for 2.5 million years. There are no luxuries: we cannot interrogate the last person seen with the victim. This is the raison d'être behind taphonomy, the study of the way in which the animals and plants fossilize and become buried, in other words the study of the grave or (*taphos*) burial.

It is not surprising that South Africa has led the way in developing taphonomy as a science. Most fossils found in this country have been in caves. This is not the case in East Africa, where most fossils have been found in the open, preserved in the sediments of dry gullies or ancient lake beds and exposed over time by natural erosion. The fact that caves feature so much in South African fossil hunting is due partly to the geology of the region —South Africa has more than its fair share of dolomitic limestone caves—and partly to the habits of carnivores, notably big cats, which usually drag their captured prey into caves.

This has been both good and bad news for paleoanthropologists. The poor record of postcranial bones, those parts of the skeleton from the neck down, is the result of the grisly way that most fossils entered the caves, dragged in the jaws of carnivores such as leopards or ancient saber-toothed cats. This is not to say that there are fewer bones in South Africa compared to the East African sites, but those that have survived are less complete than those discovered on the surface in places like Olduvai Gorge, Koobi Fora, and Hadar. Yet the bones found in caves have generally been well preserved in the darker recesses of these grottos. And in many cases, every muscle mark is preserved, which makes for better forensic diagnosis.There is less weathering. Overall, South African fossils tend to have a pristine quality about them.

Moreover, South African skulls are usually in better condition than those found at the East African sites. Consider this: A leopard dragging a poor hominid into the cave for a meal will spend more time chewing the limb bones at the expense of the skull, which is difficult to get the jaw around. And once a skull is deposited in a dolomitic cave, it is in an environment that is generally more favorable for preservation. Thus we have found beautiful specimens—the Taung Child, Mrs. Ples, and the new 573 skull from Sterkfontein—which, once retrieved from the surrounding matrix, required little, if any, reconstruction.

This is a tremendous advantage because it is often in the process of reconstruction that a scientist can insert his or her own biases into what they think the fossil should look like. This isn't limited just to endocranial volume, or brain size; it also affects the level of prognathism, or how far a face projects outward. To be fair, most scientists would not intentionally distort a skull. It is just that when there are missing parts, reconstruction experts rely on their own experience to fill the gaps, and a degree of subjectivity creeps in, which shows just how important it is to work with as complete a skeleton as possible.

A critical part of taphonomy is understanding the geological process behind cave formation. Caves are formed because erosion happens to different parts of the dolomite rock at different times. The outcrops that make up the Witwatersrand, the east-to-west ridge on which Johannesburg sits, are the geological equivalent of Swiss cheese. The holes formed underground within the water table when binding minerals such as manganese and calcium carbonate were leached out of the rocks. Geological movement caused these holes to get bigger, eventually forming caverns. Water trickling down from the surface created fissures; as erosion intensified, these became shafts. Once a shaft connected a cavern to the outside surface, this allowed for accumulating agents to act. Accumulating agents are the means by which a bone finds itself in a subterranean lair. Water is one of the more common accumulation agents, in that bones and other debris can be washed down the shaft during a rainstorm. Other common agents are predators and scavengers. A leopard or hyena could take the remains of a meal back to the cave to feed its young or to store it for eating later. Sometimes bones end up in a cave because it is a natural death trap: Animals fall through vertical shafts disguised in the undergrowth and cannot get out.

The fossilization process, then, is fairly simple. Once the bone has got into the dolomitic cave, it has a superb chance of surviving as a fossil. Calcium carbonate is eroded out of the surrounding bedrock by rainwater filtering from above and carried downward by gravity. This water, dribbling as it does through the cracks and crevices in the dolomite, picks up a greater and greater load of calcium carbonate, which remains in suspension until it comes into contact with open air, for instance the roof of a cave.

Under certain temperature conditions, some of the calcium carbonate precipitates out, forming a thin layer of lime or flowstone on the roof of the cave. If the buildup is great enough, a stalactite may form. The water droplet will continue its downward journey and eventually fall to the floor of the cave. Here more calcium carbonate will precipitate out as the water droplet seeps into the floor of the cave or runs off the surface. Again, if the precipitation is localized enough, a stalagmite may form. Any bones or other objects lying on the floor of a cave have the potential of being splashed by water droplets and becoming coated by lime. Bones that were buried within the sediments of the cave floor are constantly being saturated with precipitating calcium carbonate. Over time, the organic material within the bones is replaced by minerals, which results in the formation of fossils. Looking at these fossils, one is literally seeing a mixture of water with other substances, such as lime and sand.

When this mixture solidifies, it is called "breccia." Purer bands of lime that have not been contaminated by other debris are known as flowstones or speleothems. The end result in the cave is that there is a constant buildup of lime—from the calcium carbonate sediments, brought in either from outside or by the natural erosion and collapse of the dolomite itself—and any material that finds itself on the floor of the cave, including bones. One can see how a cave can over time fill with layers of breccia, flowstones, stalactites, and stalagmites.

Imagine a little boy filling up an empty jam jar with different kinds of colored sand. This is a fairly good analogy of how dolomitic cavities have filled up either rapidly or gradually by the forces described. The different cave fillings are called "members"; the easiest way to understand a "member" is to think of it quite literally as an "event" in the life of the cave.

The cave geology of southern Africa, however, is anything but simple. The oldest sediments are not necessarily at the bottom of the cave and the youngest sediments on top. The forces of erosion, cave collapse, and the complicated geological structure of caves have often mixed older layers with younger layers, and turned these upside down. For geologists trying to interpret the position of individual fossils within a site and the relationship to other fossils, this has been something of a nightmare and has opened them up to all sorts of criticism about inaccurate dating.

Tim Partridge of Wits University has divided Sterkfontein into six members, the oldest being labeled Member 1 and the youngest Member 6. The reason for this numbering system is that Partridge found it easier to identify the base of the cave system by its contact with bedrock. He therefore gave himself greater flexibility in numbering all younger infillings above this base. In a perfect system, the members would be one on top of the other like layers of a cake, as one episode of infilling follows another. There would, of course, have been large gaps in time where no infilling occurred because the cave entrance to the surface was closed. Theoretically, though, the layers remain in order—oldest on the bottom, youngest on the top.

But the truth of it is more complicated. At Sterkfontein, the relationship of these systems has led to a far more mixed up sedimentation than the theory would hold. Water or the roots of trees have cut into sediments, causing gaps and holes that have been filled by younger sediments. Other members have been undercut by erosion, allowing younger sediments to fill in below, so that in some cases the oldest layers sit right on top! And fossils found within sediments are often more resistant to erosion, so that the matrix that surrounds them can be washed away and replaced by material from a younger member. Furthermore, strongly consolidated breccias resembling hard concrete may decalcify over time—literally the calcium carbonate bonds break down—leaving soft sediments filled with fossil bone and stone where once there was hard rock. This leaves the potential for movement of material within the looser sediments. Time and time again, over millions of years, this process has occurred at Sterkfontein. One may well charge that to try to date fossils within the members is an impossible task that at best relies heavily on guesswork. But the situation is not that hopeless. The careful examination of the geological context of every find and the dating, not only of the member but the fossils within that member, can help us unscramble this geological omelet.

Swartkrans, just across the valley from Sterkfontein, is a classic illustration of how cave collapse can create confusion in these sites. When Bob Brain excavated the main body of sediment at Swarktrans, he found that the original cave had been undercut by a large cavity. This was partially filled by sediment before the full cave collapsed into the hollow beneath it, leaving behind part of the original sediments attached to the upper wall of the cave some ten yards higher. Known as the hanging remnant, this small outcrop represents the only "uncontaminated" sediments of the earliest infill, while all of the collapsed breccia has the potential of being contaminated by collapse debris and later erosion and redeposition. After almost a quarter of a century of work, Brain still has only a broad outline of the events leading to the formation of individual members at Swartkrans. Simple sections cannot be cut through these sites, as they are often dozens of yards deep and solidified from top to bottom. In most cases, one has to excavate almost the entire site to gain an understanding of the basics of its formation.

If the process of cave formation is complicated, so too is dating, that other tool in the evolutionary detective's bag of forensic tricks. Most of the early hominids from Sterkfontein, for example, come from Member 4, a large area of ancient infill that has been exposed to the surface through limestone mining and erosion. It is broadly divided into two areas. The so-called "Type" site, an area of hard, consolidated gray breccia, is where Broom made most of his discoveries. These are the areas that are so hard that dynamite was often used to extract the fossils. The other site is a large area of decalcified sediments often referred to as the "Swallow" hole. This softer material has been the main focus since excavations resumed at Sterkfontein in 1966. The scientific literature gives the dates for Member 4 as between 2.4 million years and 3.1 million years before present. Depending on the source, these dates have almost wholly been derived from the most common way of working out the age of South African fossil deposits: comparative faunal dating.

Comparative faunal dating is the process of comparing and contrasting fossilized animals in one site with those from other well-dated sites. The most common dating technique in the East African sites, where the sediments behave far more logically than they do in South Africa, has been through the use of either potassium-argon or argon-argon techniques, which measure the decay of radioactive potassium into inert argon. During the past five million years, East Africa has been alive with volcanic eruptions, but there have been no such events in the South African landscape. Faunal analysts comparing the fossil animals found in southern African caves with those from dated sites in East Africa are looking for individual extinct forms or species that existed for relatively short time periods and that have good chronological records in East Africa. Pigs have traditionally been the mainstay of faunal dating, because these have been some of the most extensively studied animals. But in South African cave systems, pig remains are frustratingly rare, and scientists have had to rely on other types of animals such as carnivores and antelopes.

For faunal dating to be accurate, environmental changes in East Africa, such as volcanic eruptions, would have to happen at the same time in South Africa. Specifically, faunal dating assumes that if an animal species went into extinction in East Africa it did so at the same time in South Africa, and that the southern-African and eastern-African faunal communities are broadly the same.

But both these assumptions are problematic. In Africa today one can see that species that were once widespread are now localized in one region and absent from another. This in itself may not be a permanent state of affairs, because as the climate and vegetation change, so does the distribution of animals.

Furthermore, many species in both southern and eastern Africa have evolved there, making them indigenous to those regions. For instance, the black wildebeest, springbok, and mountain zebra are endemic to southern Africa, while the Grant and Thompson's gazelles are restricted to East Africa. Confoundingly, many, if not most of the extinct species found in the South African caves, are found only in southern Africa. All this leads one to the realization that the southern and eastern African regions have to be viewed as very different eco-zones in the present and in the past. Comparative faunal dating has to be treated with caution, giving us only a very broad estimate of absolute age, if that can be regarded as a "real" date at all.

So what are we to do? There have been many attempts at obtaining absolute dates of the members or of the fossils themselves in southern Africa: Cave breccias have been analyzed to examine whether keys could be found macroscopically or microscopically, and x-ray defraction has been used to assess the elemental composition of bones within a layer in order to establish "norms." To date, success with either approach has been limited.

The most promising technique, or at least the most cited, is making use of reversals in the Earth's magnetism. At present, the planet's magnetic field is described as normal polarity—that is, the magnetic compass needle will point toward the North Pole. This hasn't always been the case. There have been periods in the past where the needle would have pointed toward the South Pole in what we would call a state of reversed polarity. These switches in the Earth's magnetism can last for tens of thousands to hundreds of thousands of years. Short reversal durations—say less than 20,000 years—are extremely difficult to document, but most geochronologists believe they have nailed down all the reversals that happened during the past five million years. Herein lies the basis for paleomagnetic dating. The process involves measuring what is called the "natural remnant magnet-ism" in flowstones and breccias and correlating these readings with sequences that have been extremely stable, most of which are obtained from deep-sea cores.

Within the Pleistocene alone—approximately the last two million years—there have been eight magnetic reversals, four normal and four reversed. They are named after either their dis-coverers, as in the Matuyama-Brunhes at 730,000 years ago, or after places, as in the Olduvai event beginning at 1,880,000 years ago and ending at 1,660,000 years ago. Paleomagnetics offers a well-documented way of assessing where one is in time by exam-ining either the natural remnant magnetism of a rock or the orientation of magnetic particles within sediment, and then assessing "normal" or "reversed" polarity. If one has a reliable sequence of reversals, and one knows approximately where one stands in time, then it's feasible to get an "absolute" date for a layer sandwiched between a series of reversals. The longer the sequence, the better the result.

In South African caves, the paleomagnetic champion has been Tim Partridge, a geographer and acting geologist at Sterkfontein for over 25 years. He's made use of this technique at all the major sites, having taken dozens of samples from breccias and flowstones within the caves. The complexity of South African cave geology, however, continues to flummox the experts, and his attempts have led to some questionable results. A single missing reversal due to erosion, or a gap in time in the infilling of the cave, can result in an event being missed; one can easily end up with an erroneous date. Furthermore, the technique relies on a broad knowledge of the age of the sediments being measured, and if one is off by several hundred thousand years—which isn't difficult when you're deal-ing with millions—one may end up comparing the wrong sequences. The obvious problem in South Africa is that we rely on comparative faunal dates to establish broad parameters, which opens up the possibility that one wrong date may lead to another.

All this is by way of explanation that the dates of all South African cave sites are necessarily broad. Therefore at Sterkfontein, a best guess for the hominid-filled Member 4 is that it is between 2.6 and 2.8 million years old. There remains a margin of error of a few hundred thousand years in either direction, and we are still unsure of the level of mixing within this member.

More recently, dating methods such as Electron Spin Resonance (ESR), Chlorine 36, and aluminum have shown greater promise for absolute dating of fossils within the caves. Right now, however, it is important to recognize that until further advances are made, in South Africa we are dealing with relative dates of fossils and not absolute dates.

Anatomy, taphonomy, and dating are scientific processes that help us analyze what a fossil is, how it got into the cave, and how long it has been there. It might also help explain the nature of an animal's death. This helps reconstruct the objective truth of a case. To return to the detective analogy, these techniques will tell us that an underworld mobster died through drowning because he was thrown off the Hudson River bridge into the water below, and the concrete in which his ankles were set prevented him from swimming to the surface. We can surmise that this happened on February 10, as that is when his watch stopped working. As useful as this information is in preparing a case, it tells us nothing of his lifestyle, what activities he had been engaged in, and why he should have ended up with his feet in concrete blocks in the Hudson River.

Only when hominids began manufacturing artifacts—the earliest being 2.5-million-year-old stone-age tools found in Ethiopia—did we have concrete evidence to use in understanding early hominid lifestyles. To get a glimpse into the minds of their ancestors we have to turn to the circumstantial evidence of primate behavior. To watch a chimp in a zoo or a circus—or in the wild—is to look back in time to when our hominid ancestors parted ways with a lineage of chimpanzee ancestors. It is important to remember that we did not evolve from chimps, but from an ancestral form of African ape. That 3 percent genetic difference between chimps and humans today is the result of several million years of evolution, leading to a conceptual gulf between us that is as wide as the universe. But despite this gap, we recognize something of ourselves in chimpanzees, which is why they are so effectively used as a human foil in films and advertising.

Because chimps are our closest cousins, many scientists have studied their behavior to try to understand the psychology of early hominids, a gap that can't be filled merely by looking at fossil bones and understanding anatomical relationships. Of course, a great deal of latitude has to be allowed in doing so; such an exercise can at best only be circumstantial because chimps, too, have evolved from that common ancestor. In doing so, one could do worse than spend a couple of hours in the company of my good friend Jane Goodall, who is probably the best authority in the world today on chimp behavior. There are numerous subspecies of chimps living in a broad central African band from Guinea, in the west, to Tanzania, in the east. The common chimpanzee (*Pan troglodytes*) is found in a wide range of habitats, including tropical forest and broken woodland. These chimps generally build their nests in trees and depend mostly on fruit for their diet. They exist in fluid social groupings, which are determined by food gathering. Females and juveniles often split off from the group to find their own food, while males tend to stick together in tight alliances.

When Goodall first began studying chimpanzees in central Africa in 1960, Louis Leakey warned her that she was taking on a ten-year project. She replied with a laugh that she thought three years would be sufficient. Now 40 years later, her work has revealed to us the complexities of the chimpanzee world, and in so doing has provided us with a remarkable insight into the likely behavior of our earliest ancestors.

"Had my colleagues and I stopped after ten years, we would have been left with the impression that chimpanzees are far more peaceable than humans," she told NATIONAL GEOGRAPHIC. In fact chimps in many ways mirror the emotions of humans. Like people they can be "kind or cruel, caring or cold, thoughtful or stupid." When she first started at Gombe Stream National Park in present-day Tanzania, she believed that "chimps were nicer than we are." But time has revealed that they are not. "They can be just as awful."

Goodall was shocked when in 1974 one of the four groups of chimpanzees she was studying at Gombe, a group known as the Kasakela community, launched an attack on a splinter faction of chimpanzees. The war lasted four years and ended only when every member of the breakaway group—seven males, three adult females and their young—had been killed. Eyewitnesses described how the chimps moved silently and stealthily through the forests in single file, hair bristling with excitement before attacking their victims aggressively with their teeth and fists. Researchers witnessed five such forays, which included female chimps killing and eating the infants of their enemies. Nothing like this had been seen before.

This is not to say that we evolved from a group of ancient killer apes. This is simplistic and misleading because it does not take into account the complexity of primate relationships in which aggression is but one characteristic. Goodall believes that aggression among primates is directly linked to dominance. Primates live mostly in social groupings with distinctive hierarchies. By being aggressive, or at least displaying these tendencies, high-ranking individuals can displace their lower-ranked colleagues from food, water, a sleeping site, or even a social partner and a mate. Although there is a complicated relationship between dominance, subordination, and kinship, a high rank often leads to breeding success because the alpha males usually have priority over females in estrus.

Among Goodall's other insights during her years of studying chimps is that these animals are not complete vegetarians as was previously believed. In her first year of study in Gombe, she witnessed a chimp she named David Greybeard, the first of the animals to accept her presence, eating a baby bushpig. She also watched David Greybeard trimming a wide grass blade to probe a termite nest, extracting the insects and eating them. This was the first time nonhumans had been seen manufacturing tools and using them in the wild. The Mitumba chimps were the only group at Gombe that regularly used twigs and blades of grass to catch termites or ants. Then, in 1994, when a Mitumba chimp joined the Kasakela community, Jane witnessed the transfer of technology as the twig skills were passed on to the new group. In more recent years, studies of chimpanzee groups around Africa have shown that complex tool manufacture and use are not that uncommon. In fact, over three dozen different "cultures" of tool use are now known to exist. The most ingenious are probably the wooden "hammers" used by the chimp community in the Tai National Park, Ivory Coast, which are used to crack open nuts to get at the protein-rich food inside.

Studies involving chimps in captivity have also revealed that they have a consciousness about themselves that appears to be absent in other primates. The ability to conceive of self indicates a capacity for abstract thought, which in itself led to the first rudimentary technological innovations, such as the making of stone tools.

A series of mirror experiments with chimps were first conducted in the 1960s and 1970s to gauge the capacity of their consciousness. When these animals studied their reflections, their first responses were social, perceiving their reflections as if they were seeing another chimp—in other words, something "other" rather than "self." They gradually learned how to relate to their images as their own. In a now famous experiment, chimps that had experience with mirrors were anesthetized and a blob of red dye placed on their foreheads. When they awoke and were confronted by their reflections, they immediately noticed the red spots and tried to groom them off. Chimpanzees that had not previously handled mirrors could not use the reflection to remove the spot, while gorillas could not recognize themselves in mirrors at all and were oblivious to the red spot treatment.

Another set of experiments involved the sharing of knowledge. Wild chimpanzee males are known to call others when they come across fruiting trees. Captive chimps were taken individually into their enclosures and in a roundabout way shown the location of a box of hidden food. They were then removed from the site and subsequently returned to the site with the rest of their social partners. Those chimps that knew where the food was took the others directly to the box. Goodall's experience shows, however, that consistency among individuals should be not expected. One chimp was shown a box with food and another box with a snake. When he was introduced to the enclosure with the rest of the group, he took them to the box with the snake. They took fright and retreated, whereupon he went to the box containing the food and ate it alone. The capacity to deceive may lie as deep within our primal ancestry as the capacity to share knowledge.

But one should be cautious if speculating about the extent to which ancient primate behavior genetically resides within us, determining modern human behavior. It can be tempting to use examples like the incest taboo that exists within chimp group relationships to explain why it also exists within human society, but there may not necessarily be a correlation. Primates are social animals and the rules of group behavior are generally a product of a social environment. That social environment is, however, ultimately shaped by a biological one. It is reasonable to speculate that the earliest hominids lived in territorially defined social groupings with male-dominant hierarchies, and food gathering shaped the group behavior.

This fossil forensic aspect of paleoanthropology is one of my favorite parts of being a scientist in this field. Paleoanthropology combines the best of good detective work with the thrill of treasure hunting, or searching for some of the rarest objects on Earth. Like any good detective novel, a case is solved when the final piece of the puzzle fits into place. In the case of real-life detective work, the mystery may never be solved, leaving only speculation.

CHAPTER 6

SLAUGHTERING
SACRED COWS

⌐⌐

T he conservatism that marked South African politics in the second half of the 20th century was also evident in its science. From its position during the thirties and forties as the front-runner in the search for human origins, South Africa slid abysmally during the sixties and seventies into an inward-looking nation. Intellectually stifled by apartheid and shunned by the rest of the world, the country took on a garrison mentality. In some ways the achievements of the first half of the century—the discovery of Taung and Mrs. Ples and the confirmation that human ancestry was firmly located in Africa—contributed to a sense that there was nothing new left to discover, that all the science could do now was fill in the gaps. There was also the towering presence of Professor Tobias, whose personality ultimately cast a long shadow over South African paleontology.

When I arrived in 1989, one of the first things that I had been told by almost every paleoanthropologist I met was that if I intended to find hominids in a cave site in South Africa, I first needed to find baboons. This had been the dogma since Dart had discovered the Taung Child among scores of small monkey skulls at the Buxton limeworks, and it had been reinforced by Broom, who actively sought out fossil sites, such as Sterkfontein, that were rich in primate fossils. Ever since, this approach had been carried on through the ranks of paleontologists. The reasoning was sound. It was based on the assumption that whatever predators had collected baboons would have also favored hominids. Every site that had revealed fossil monkeys—Taung, Sterkfontein, Kromdraai, Makapansgat, and then Swartkrans—had initially been identified as containing abundant baboon assemblages. All had within a short period of time produced hominids. I had a strong belief, however, that this thinking was limited, that it had developed into a circular set of assumptions that were not advancing the science. I suppose I also had the advantage of being an outsider whose thinking hadn't been shaped by the conformity that had developed in paleontology during the Tobias era. I felt sure that if one actively avoided looking at sites that didn't have baboons, one was certainly not going to find any more hominids; plus, it just didn't make ecological sense. If hominids were present in the past in any significant numbers, then random accumulators such as scavenging hyenas, and even porcupines, should pick them up, even if only in low numbers. I was confident that if we didn't look into these other sites then we were never going to advance our understanding of early hominids. I had therefore spent the last two years on an opposing track, patiently looking for a site that didn't have abundant primates yet represented enough of the animals living in the ancient Witwatersrand environment.

The immediate consequence of this decision was that I did a lot of walking. At every opportunity, I trawled the dolomitic hills and valleys that lay beyond the peri-urban sprawl of Johannesburg's West Rand. I searched the confounding maze of riverine forest kloofs of the Magaliesberg Range that lay between Johannesburg and Sun City. I trekked across the broken hills that surrounded the Hartebeespoort Dam, west of Pretoria, including an area called Pelindaba, which used to be off-limits to civilians because it housed the former white government's top secret nuclear weapons research facility. I also came across the abandoned "death squad" farm, Vlakplaas, tucked away behind a wall of willow trees and a bubbling brook. This was where the security police of the former regime planned their covert operations against people involved in the anti-apartheid movement. It was also suitably isolated for them to interrogate and torture ANC guerrillas that they'd captured during the height of South Africa's recent violent past. An eerie atmosphere pervaded Vlakplaas, which on the face of it was just an old ramshackle farmhouse surrounded by a few magnificent palm trees and a number of outbuildings, cynically sign-posted as "guest" cottages, which on closer inspection were heavily barred and were probably interrogation cells. "You'll probably find a few hominids buried there," a journalist once commented, with the dark humor that South Africans often resort to. "But I don't think they'll be of the antiquity you're looking for."

During my walkabouts I developed an intimate knowledge of the geology, flora, and fauna of the area and was surprised at the amount of wild animals that existed so close to human habitation. I soon learned to look out among the boulders for trees like the wild olive and white stinkwood (*Celtis africana)*, which because of their love of lime soils and deep-rooting systems were often an indication that a subterranean cavern lay beneath them. On more than one occasion I crawled into caves inhabited by leopards, and on several occasions came face to face with these beautiful, deadly animals. After a few close calls, I kept a cautious eye out for them. I certainly didn't want to suffer the same fate that had befallen many of our early ancestors, whose remains I was looking for in these hills. Then I finally hit upon the perfect site. Gladysvale. Ironically, it came to me in my lab at Wits rather than during my outdoor excursions.

In early 1991, a block of brown breccia was brought into the anatomy department at Wits University. I just happened to walk into the main office to find this stone matrix lying on the desk of the departmental secretary Heather White. She told me that a little old man had dropped it off, saying that he'd found it in a cave and that it may be of interest to us. My curiosity was immediately aroused. I had a careful look at it and was astounded to make out that it contained the crushed but complete skull of an extinct saber-toothed cat. This was all the more astonishing because complete specimens of carnivores are extraordinarily rare, perhaps even more so than hominid skulls. The man had left a note with his name and phone number on it. I called him that afternoon.

Miner's dumps, full of fossilized remains, are clearly visible near the main entrance to the cave at Gladysvale.

At first he was hesitant to answer my questions. Then he became deliberately evasive and stubbornly refused to reveal the location of the cave in which he'd found the fossil. I tried to impress upon him how important this find could be, and that unless we knew the context in which it had been discovered, we would not be able to do justice to this specimen. He still wouldn't be drawn into discussing anything further on the phone, and I had to persuade him to give me his address so that we could meet face to face. He grudgingly agreed and gave me the details of where he lived. A couple of days later, with the rock-encrusted saber-toothed skull on the seat beside me, I drove to his home some 20 miles from Johannesburg.

In his sitting room, with a cup of tea next to me and the skull in my lap, I pointed out all the features that had convinced me that this was indeed a rare find. I implored him yet again for the location of the cave from where the block had originated. Eventually I squeezed it out of him.

"Gladysvale," he muttered reluctantly.

The name rang a bell. I'd heard it before but couldn't place it.

"Where is it?" I asked.

"Not far from here, I think," he answered.

Hope rose in me. "Can you show me on a map?" I asked, expecting a positive reply.

"I don't think so," he said, looking away from me.

"You see, I found that fossil over 40 years ago—" he paused before continuing— "and my brother and I stole it."

A fossil thief. No wonder he had been so reticent. He must have worried that I would turn him over to the authorities. I made light of his antiquarian endeavors, but the removal of fossils is a criminal offense, a violation of the National Monuments Council Act of South Africa. Thieving is a problem as old as the profession itself, and deeply frustrating for scientists because even if someone is trying to be helpful by bringing in a fossil, much of its value is destroyed when it is taken out of its original context. Nevertheless, I assured my host that his cooperation would not get him into trouble. He visibly relaxed and told me that 40 years ago he and his brother had visited a private game park "near Krugersdorp." They'd left their car and explored a cave, where they noticed a stone block because of the large fossilized teeth protruding from it. They hid it behind the seat of their car and drove out of the reserve, telling no one what they'd found. For the next four decades, he used this extremely valuable specimen as a doorstop in his modest home. Now, all these years later, for some reason he couldn't really explain, he'd decided to turn it over to the university. I was reminded of the Taung skull, which would have ended up as a paperweight on a lime-quarry manager's desk back in the twenties had a geology professor not intervened.

As I drove back to Johannesburg I considered my luck. I had repossessed a stolen fossil that was invaluable in itself, and which could also be a clue to finding a hominid site. But my hopes of finding the exact location had been dashed. "Near Krugersdorp" was so general as to be useless. It could have been in any one of the hundreds of caves in the vast area that I'd already been searching. But I did have a name: Gladysvale. Where on Earth had I come across it? Back at the lab I stared at the collection of classic anthropology texts that I'd built up since I'd arrived in South Africa. Where could I find it? I eventually decided to start at the beginning and work my way forward. I picked up the classic seminal monograph by Robert Broom, *The South African Fossil Ape-Men: The Australopithecinae*, published in 1948. Opening the book, I started to flip through the pages when I read the following paragraph:

> Then Mr. G. van Son, our entomologist, told me of a large cave at Gladysvale about 15 miles north of Krugersdorp, where he had once seen what looked like a human jaw in the breccia in the wall of the cave. Unfortunately he had no implements with him at the time, and on his returning a little later found that someone else must have seen it, and removed it.

Bingo! Not only had I found what Gladysvale was, it might even have hominids! I was overjoyed, but the feeling was short-lived. After spending five more hours scouring every major text on South African sites, I could not find mention of the exact location of Gladysvale. I also knew that it couldn't be "15 miles north of Krugersdorp" as Broom had said, because that area had the wrong kind of bedrock in which fossils would be found. That evening I went home excited that I was on to something, but frustrated in that it remained beyond my grasp. For the next two weeks, I hunted unsuccessfully for references on Gladysvale. One person who would certainly have known was Bob Brain at the Transvaal Museum, but the receptionist there told me he was in the field and out of contact. What about Professor Tobias? Unfortunately, he was away on one of his frequent overseas lecture trips, and none of the younger scientists had any idea where I might find this cave. I was on the verge of giving up on Gladysvale when my luck changed

At the time, one of my duties as a graduate student was to assist in supervising bachelor of science students with their class projects. One young man had decided to undertake a historical project on the Sterkfontein site and made an appointment to see me to discuss his progress. When he came in, we sat for about half an hour discussing his work and I read excerpts from his draft manuscript. As I read through his work, I became suspicious of his references. It looked as if he had gleaned much of his information from an encyclopedia.

"Where did you source this?" I waved the manuscript accusingly at him.

"From here," he said defensively, holding up a large red volume of the *Standard Encyclopedia of Southern Africa*, which he took from his case. "I borrowed it from Professor Tobias's office," he added, as if that in itself would allay any concerns I might have entertained.

"Well," I said, in my best stern professor-to-be voice, "you should know that we don't use encyclopedias in university research. Find the original texts." To reinforce my point I confiscated the book from him and sent him back to the medical school library.

I walked down the corridor to replace the book in Tobias's office, and as my steps echoed down the passageway, a thought struck me. If Sterkfontein was in the *Standard Encyclopedia*, what chance was there that Gladysvale might be, too. As I slid the "S" volume back into its space on the professor's shelf, I looked over at the "G" volume. Taking it down, I flipped through to GLA, and there it was: Gladysvale. How obvious, I thought. If you want to research something, find an encyclopedia! I read through the short couple of paragraphs, which outlined a little more history than I had, including a visit to the site by an American team that came to South Africa on an unsuccessful hunt for hominid fossils in 1947, the Camp-Peabody expedition. Then I read, "in 1946 a student expedition led by the author...." I didn't need to read further; the style of writing was already immediately recognizable. Flipping to the author list just confirmed my hunch. The text had been written by Phillip Tobias. What do you know. I decided then and there to allow my students to use encyclopedias in their research if they so wished.

Tobias returned the next day. I waited until he was settled in his office and had dealt with the initial paperwork that coagulates like grease in a sink on an academic's desk while he or she's away. Then, armed with the saber-tooth fossil and my references, I walked into his office and briefly outlined the story of my discovery and the search before finally asking the question.

"Do you know where Gladysvale is?"

"Of course I do," he replied, as if he'd been asked whether he'd ever had an interest in fossils. Then a twinkle came into his eye. "What's more, I can get the owner to show you. His name is John Nash, and he called me a few weeks ago to tell me he had found a cave on his property. I told him it was Gladysvale, but that there wasn't much there as we had already looked in 1946. Anyway, if you want his number, here it is." He tossed a small memo pad across his desk.

A week later, I visited the site with John Nash and fell in love with it. Gladysvale cave lies in the middle of an immense private game reserve, tucked away near a government satellite tracking station in the foothills of the Magaliesberg Range, northwest of Krugersdorp. It's actually very close to Sterkfontein, which shouldn't have surprised me. It has a primal feel about it. There are dozens of species of antelope, as well as giraffe, zebra, leopard, and hyena. It had everything I wanted: plenty of fossils, a refreshing lack of primates, and was extensive enough to give good samples. The reserve manager, Joe DeBeer, told me two things on that trip: that John Nash's son, Paul, had once held the 100-meter sprint record in South African athletics, and that a geologist from the Geological Survey had already studied some of the breccia dumps and that I should contact him. His name was Andre Keyser.

I called Keyser on the phone the next day.

"Sure I've got some fossils from Gladysvale," he said in his scratchy but friendly smoker's voice. "And if you are interested in excavating, let's meet at the site and I'll show you the underground."

We rendezvoused at the old Gladysvale farmhouse and drove through the golden grass and broken woodland of the reserve to the cave. Keyser also revealed that he was an amateur spelunker and would take me down into the deepest part of the cave, where the fossils were even more abundant. He showed me a publication that he and another amateur enthusiast, Jacques Martini, had compiled a few years before, surveying the underground system using a tape and compass as part of a mapping project of the caves of the Nash reserve. I was impressed by his confidence and readily joined him as we scaled down a slim rope into darkness. As we descended along a steep 45-degree slope, the light from our head lamps illuminated a giant talus cone of debris spilling down into a large cavern. When we reached the bottom, some 50 yards down, bats fluttered around our heads as I gazed at the enormous cavity. Looking at the walls and ceiling, I could see white specks of bone sticking from the rock. Miners had been at work here, and debris and rubble lay piled everywhere.

"Fantastic," I said to Keyser.

"You haven't seen anything yet," he replied.

He was right. We crossed one of the largest piles of rubble and crawled under a low ledge, maybe a foot and a half high. As I struggled on my back through this confine, I looked up to see inches from my nose one of the densest bone assemblages I had ever come across. Crawling to the end of this passage, we started climbing down a three-foot-wide vertical shaft. Several minutes later, and 15 yards down, we had reached the bottom, a wonderful, awe-inspiring chamber maybe five by five yards across, glittering with crystalline structures and bones protruding from every nook and cranny.

"So you want to work here?" Once again, the sight of South African fossils had taken my words away.

Within two months I had organized a field excursion to Gladysvale with some of my students from Wits. "You're not going to find hominids there," my university colleagues warned me when I told them of my intention to dig Gladysvale. "That whole area was thoroughly searched years ago; there's nothing left to find." Eighteen days into the dig, we hit pay dirt.

"Do you think this is a tooth?"

The innocuous-sounding question came from Michelle Erasmus, a B.Sc. student, who held up a small lump of pink breccia in the bushveld afternoon light. She'd found it protruding from the side of a rock almost three weeks into our excavations at Gladysvale. I squinted at the rounded piece of enamel, and my mouth suddenly went dry. Could it be? Even across the two yards or so that separated us, I could see that it had the right curvature and bulbousness, and judging by the pearly nature of the side I was looking at, the enamel looked thickened. Not only was it a tooth, but a hominid one at that.

"It certainly is. In fact I think it's a hominid."

These words brought an instant halt to all the activity around me. The fieldworkers and students who moments before had been sorting breccia clustered around Michelle and me as we peered at this little tooth sticking out of the rock matrix. There was a collective chorus of interest and disbelief.

"It can't be."

"Wow, that looks good."

"It's not a tooth."

"Why do you think so?"

I barely heard the tirade of comments and questions as I went through a mental checklist to overcome the doubts in my mind, despite what my eyes were seeing. I stared at the quarter inch of enamel. I was right in my first impression. It was a hominid, a hominid from Gladysvale. I hugged Michelle who was staring in disbelief at what she'd found. "Congratulations, you've made history."

This australopithecine premolar made Gladysvale the first early hominid site to be discovered in southern Africa in just over 44 years. The last time a new site had been discovered was in 1948 when Robert Broom came upon Swartkrans. Within a couple of days, we had a second, discovered by Joe DeBeer, Nash's game manager. Analysis would reveal that they once belonged to a gracile australopithecine, probably *africanus*. When they were announced in the *Journal of Human Evolution*, I was astounded at the response from around the world. You would have thought we had found a whole skeleton. Newspapers, radio, and even a small piece in National Geographic taught me that the world was hungry for good news from the fossil caves of South Africa. This country's paleontological renaissance was just beginning. Gladysvale also taught me an important lesson. Just because everybody believes that something is true doesn't make it so. There were more fossils, fossil sites, and discoveries to be made in South Africa than one could imagine. One sacred cow had just been brought to the slaughter. How many others could be?

One of paleoanthropology's unsolved mysteries was what had killed the Taung Child. The question had vexed Raymond Dart to his dying day, and was one that he never answered satisfactorily. The conventional wisdom was that a leopard had killed the little australopithecine child. This theory was given credence by one of South Africa's most respected scientists, Bob Brain, who is a personal friend and mentor. Brain, who recently retired as the head of the Transvaal Museum, pioneered taphonomy—the study of graves—in South Africa. Universally respected, his quiet and gentlemanly demeanor hides an aggressive scientific brain that is probably unrivaled in the field. He began his career as a geologist, studying the way in which the ape-man caves had been formed. His inquiring mind soon drew him to the additional and complex problem of what had collected the bones in the cave.

During the mid-fifties, Raymond Dart had developed a theory about the culture of the australopithecines, which he believed were the main cave accumulators. Based on the fossils recovered from Makapansgat, Dart surmised that their fragmentary nature was due to what he called Osteo-Dento-Keratic (ODK) culture, a term that has taxed the tongues of anthropology students possibly more than their minds. ODK literally means bone, tooth, and horn culture. Dart had identified the Makapansgat ape-men as a new gracile species, *Australopithecus prometheus,* on the mistaken assumption that *prometheus* had the ability to use fire. *Prometheus* was later reclassified as *A. africanus,* when the bones were found to have been stained by manganese and not burned by fire at all.

Nonetheless, at the time Dart concluded that not only had *prometheus* smashed and butchered animals using bone and horn tools but that he was essentially a killer ape and cannibal. Dart essentially saw humanity as inherently aggressive and bloodthirsty, a view probably influenced by his gruesome experience as a medical orderly in the First World War. He speculated that this behavior was already latent in the australopithecines, writing later that the Makapansgat ape-men were "flesh-hunters" whose violence led inevitably to the "blood-spattered, slaughter-gutted archives of human history." American popular-science writer Robert Ardrey enthusiastically absorbed his ideas, and maintained that humanity's bloody birth and primal aggression has been passed on genetically through the ages. His book, *African Genesis*, was as popular as it was scientifically misleading and went on to become a bestseller in the 1960s. Dart's theories also were picked up cinematically. Most movie fans will recall that brilliant opening sequence in Stanley Kubrick's *2001: A Space Odyssey* in which a bloody bone tool is thrown high into the air by an ape-man and then morphs into an intergalactic spaceship. Today ODK is viewed somewhat whimsically as an aberration of scientific thought that has severely undermined the reputation that Dart established with the Taung discovery.

At the time Bob Brain was already suspecting that things weren't quite as Dart had suggested. Brain had vast experience in examining caves and their contents. He felt strongly that hyenas, leopards, and even porcupines were probably the main culprits in collecting fossil bones in caves rather than the more colorful, bloodthirsty killer ape-man. He undertook a methodical and deliberate study of collections of bones from almost every living and dead potential accumulator he could find. Over the next several decades, he conducted innumerable experiments, observing how animals ate, chewed, gnawed, collected, and interacted with individual bones and complete skeletons. At the same time, he also ran the excavations at Swartkrans for a 26-year period, using the site to test his hypotheses. In the process, he set the standard for excellence in site excavation, using levels of precision in recording that were considered unnecessary by most of his colleagues but in hindsight have proved invaluable.

In the course of his experiments Brain uncovered several hundred fossil hominids, mainly robust australopithecines and even some *Homo erectus* fossils. One of his most significant discoveries was the first use of domesticated fire, which he found at

Swartkrans. Thin layers of burned material were found in the cave walls where the floor level would have been 1.1 million years ago. Their burning patterning suggested that they were not random but deliberate, making Swartkrans the site of the earliest documented evidence of the controlled use of fire. Previously, the earliest use of fire was believed to have been about 450,000 years ago at the Zhoukoudian Caves near Beijing, the home of Peking Man. Brain believes that the *erectus* hominids he found at Swartkrans did not have the technology to make fire themselves, but had probably found a way of keeping a natural bushfire alive in their caves. Fire would not have been an unknown phenomenon to early hominids. They would have experienced bushfires created mostly by lightning strikes, a common cause of runaway fires in Africa today.

The Swartkrans hominids then were the real prometheans rather than the ape-men of Makapansgat.

Brain addressed a fundamental issue about the early hominids. Were they the bloodthirsty hunters of Dart's world, or did they occupy a different niche in the African environment? It was a critical question, one that was central to our understanding of why we are the way we are. Bob Brain published his findings in a monumental book titled *The Hunters or the Hunted?* and conclusively demonstrated that it was not the ape-men that had accumulated the bones of the South African caves but carnivores and scavengers, particularly hyenas, saber-toothed cats, leopards, and porcupines. He found that the damaged bones of a young ape-man child were not caused by the bludgeoning of cannibals, as Dart had suggested, but were the result of hyena gnawing. Leopards were particularly important to Brain, who demonstrated that they were the most probable collecting agents of the ape-men at Swartkrans. He even had the proverbial "smoking gun" to prove it. Brain discovered a child's skull in the assemblage with two neat round holes punched through its head. When I first saw this specimen, I could almost imagine Raymond

Dart lifting up an antelope horn and demonstrating ODK culture by driving the point into the skull, not once but twice just to make sure. However, Brain had found the real murder weapon not far from this child's skull. It wasn't the horn core of an antelope, but the fangs of a large leopard that had obviously preyed on the child. The leopard's canines fitted exactly into the two holes in the skull, matching the spacing and diameter perfectly.

Brain made a convincing case that our ancestors were the hunted rather than the hunters. His argument was so forceful that it sent ODK apologists scuttling almost overnight. Brain believed that leopards did not necessarily take their prey into the caves themselves but would often drag their victims into the trees growing at the entrance to these rock cavities. There they would chew on the bones, which would tumble down into the caves where they would become fossilized. The strength of this imagery is such that it has become iconographic. Almost every textbook or popular science book portrays the image of a poor australopithecine dangling from a limb of a tree with a proud leopard crouching over it.

Soon after I was first introduced to Gladysvale, I came across my own leopard lair, not far from the Gladysvale cave. It was an extraordinary sight to behold, containing the remains of no less than 14 animals, including the complete skeleton of an eland, the largest antelope in the world. A female leopard that proved to be a prodigious collector of animals had occupied the lair, creating one of the densest bone accumulations I'd ever come across. What made this assemblage particularly significant was that it was in a dolomitic cave in the very valleys where our ancestors used to roam. In fact it's as close to an actual comparison of a Plio-Pleistocene cave that we were ever likely to get. There were two aspects in particular that interested me. First, how would the fossilization process occur, if it did at all? Second, I was keenly aware that the leopard had accumulated the bones *in* the cave rather than in a tree in the entrance of the cave. Did this mean that the old "leopard in the tree hypothesis" put forward by Bob Brain might actually be flawed? Along with Darryl DeRuiter, a graduate student of mine, we spent seven years conducting a taphonomic experiment, comparing this assemblage with other caves in the area. We found a repeating pattern. If the leopards were anywhere in the vicinity of a cave when they made a kill, they would drag the carcass into the cave rather than use trees that were much closer. During the years of observation, we never documented a single case of a leopard using a tree in the entrance of a cave as a cache. Instead, they would always use the recesses of the cave itself.

The distribution of the assemblage, random or deliberate, may have implications for how we interpret the evidence emerging from a site. After monitoring the assemblage and other caves in the reserve for seven years, we felt we had a good case for arguing against the leopard-in-the-tree idea and instead proposed the "leopard-in-the-cave" hypothesis.

Gladysvale was a fine testing ground for Bob Brain's methods of taphonomy. The accumulations here were denser and more numerous than at other sites, but they were also fragmentary. Bob had developed a system of analyzing antelopes by body size, so that by classing groups of animals by weight and bone size together, one could get an idea of the size of the accumulating predator, or the size of preferential prey of animals. I wanted to apply this method at Gladysvale, not only to the antelopes but also to the carnivores, pigs, horses, monkeys, and even the small and medium-size animals like hyraxes and hares. I took the adult body weights and juvenile body weights of animals that I had found at Gladysvale that had living relatives. If the animal had no living counterpart, I would use an animal closely related in size and behavior to obtain the estimate. I then divided them into large, medium, and small-size animals. I did the same for numerous modern carnivore and scavenger collections. Combined with the breakage patterns of the bones, the results allowed me to make an educated guess as to whether the assemblage had been collected by a leopard-size cat, or a spotted hyena-like animal, a slightly smaller brown hyena, or a lynx or whatever.

I was so pleased with the results at Gladysvale, which incidentally seem to be largely collected by hyenas, that I applied the same technique to all of the other hominid sites. Not surprisingly, Swartkrans came out with a mixture of collecting agents, and as predicted, the dominant collectors were large cats and hyenas. Sterkfontein looked like a large-cat collection; at Makapansgat and Kromdraai, the accumulators were hyenas.

The enigma was Taung. It didn't fit any accumulating agent for which I had data. The collection was composed almost wholly of small monkeys, hyraxes, and tortoises and there was even the odd eggshell or two. The only bovid remains were juvenile or a single limb bone or two of something larger. I wasn't the first to see this oddity. Dart had, of course, accused ape-men of assembling Taung as a "kitchen midden." Bob Brain had looked at Taung, noted its unusual assemblage of small animals, but suggested that in absence of a better answer, they were probably collected by a leopard. Looking at my crude data, I didn't think that was possible. Wouldn't a leopard crush the small hyrax skulls? And what about the tortoises? I had heard of hyenas collecting tortoises but not leopard.

At Gladysvale, I have a favorite rock that I like to visit when the day's excavations have ended, a place where I often sit and imagine what the world must have been like for our distant ancestors. Gladysvale is a particularly appropriate place for this type of reflection since the cave not only has fossil hominids but it lies within a pristine, game-filled environment. From my perch, I can look down the valley for about a mile to a dense forested area that hides a crystal-clear river that flows all year-round. The soft undulating rhythm of the rushing stream blends with the noise of birds in the forest to create the effect of a gentle bushveld orchestra, in which the haunting calls of the hornbills and the raucous cry of the gray lourie alternate as the lead instruments. It's not hard to imagine the source of inspiration for the first human musicians. Beyond the river, the ground rises steeply into long rolling grasslands where, on any given day, with a good pair of binoculars, you can watch antelope, zebra, and giraffe feeding on the slopes.

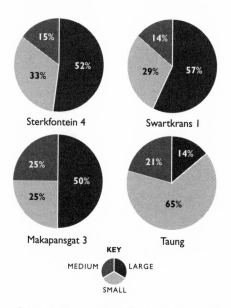

The distribution of large, medium, and small animals at several cave sites in South Africa yields clues. An extremely high concentration of small animal fossils found at Taung first led the author to suspect a different agent of accumulation at this site.

In early 1993, I found myself on this rock, puzzling over the Taung accumulation and wondering what had killed that australopithecine child. Across the valley I noticed a troop of vervet monkeys moving carelessly over the warm rocks on the lower slopes of the hills, sunning themselves in the late afternoon light. Dusk was an hour away and the monkeys were clearly in no hurry to make their night shelter in the protective thicket of thorn trees beyond the large chert boulders. The younger vervets scampered insolently between the rock figs, their humanlike shrieks punctuating the bushveld air as they played tag with the ageless abandonment of children. The older monkeys groomed each other in a ritual of communal bonding, plucking and eating the lice from each other's bodies. Occasionally, their black faces would turn to the shallow valley behind them. Unperturbed by my presence, they lazily kept a watchful eye on me. They could see that I was not armed, and their experience in the surrounding farmlands had taught them that a man without a gun was not a man worth fearing. To me they were objects of primordial fascination, a welcome distraction from my taphonomic mind puzzle. From time to time I'd look at the monkeys and contemplate the distance between us. Two hundred yards and 30 million years. They are the beginning of the journey and I am the end.

An agitated commotion from the rocks distracted me. Almost immediately I sensed the object of the monkeys' anxiety as a large shadow passed me by. I glanced up in time to see a magnificent black eagle gliding past toward the forest canopy, like some modern warplane flying low to avoid the enemy radar. The bird was close enough for me to discern the distinct white V-shape across the top of its two-yard wingspan as it banked slightly in the breeze, and the stark yellow of its beak and talons in marked contrast to its powerful dark body. The black eagle is one of the most imposing African birds of prey, and I was so caught up in this vision of airborne power and grace that I almost missed the second black eagle hurtling toward the trees from the opposite end of the valley. In that instant, I realized that the first bird was acting as a decoy, distracting the monkeys away from their mates. Before this thought had fully slipped into my comprehension, the second raptor swooped into the trees, and in an act of terrifying grace and ease, knocked an unsuspecting monkey out of the branches, sending it howling and tumbling onto the rocks below.

The monkey survived the fall but lost its bearings. That split second of disorientation, though, cost it its life as the decoy eagle, which had spiraled out of sight during the maneuver, swept down and impaled the little gray body in its talons, killing it instantly. I was stunned by the intensity of the experience. The monkeys, so clamorous just moments before, had disappeared into the bush, leaving a shocked silence in their wake. For a single moment there was silence and the suspension of motion. The two eagles, one poised on the body of the dead monkey and the other perched on a tree above it, turned to look at me. Then, with a lack of urgency that bordered on arrogance, the first eagle flapped its wings, once, twice, and then took off, clutching the lifeless vervet firmly by the head, winging its way back across the valley floor from where it had come barely 60 seconds before. The second eagle joined it within seconds, and the two birds disappeared from sight behind the dolomitic hills that were now consumed in the evening shadows.

I felt slightly dizzy, horrified at the unexpected violence of the drama I'd witnessed and by the speed with which death had overtaken the monkeys. I was also thrilled to have witnessed a rare kill in the African bush. Somehow, through the jumble of unconnected thoughts and emotions, I centered on the image of Taung. Could the accumulator have been a bird of prey? What a fantastic thought.

I raced back to the library at Wits to find all I could on African raptors. I realized immediately that I was on to something. Almost every book I opened gave me lists of prey that mimicked the species list from Taung. Hyraxes were the preferred prey of black eagles, I read. They are often seen to collect tortoises. They had even been seen to take primates! I found out for the first time that these great birds were prodigious bone accumulators, not deliberately but as a result of their bringing back prey to the nest to feed their ever-hungry young.

The next day found me back at Nash's farm, working my way along the base of a cliff where I'd seen the black eagles return after their monkey raid. Crossing through a thicket, I could see the enormous nest about 30 yards above me tucked against the sheer face of the rock. I glanced around and saw I had to be on the right track. There were bones everywhere. They ranged in size from small hyrax skulls to the bones of a small antelope. And nearly all of the material had punctures or showed strange V-shaped cuts from the beaks of the birds. Within a few minutes, I had collected a small sample of material to take back to the lab. Pleased at the variety of bones but disappointed by the lack of primates, I began walking away from the cliff face. As I stepped through the undergrowth, my eyes were drawn to a small white object about the size of a baseball hidden under some leaves. I leaned down to pick it up. In my hand was the reward: the top of the cranium of a juvenile baboon with a large depressed puncture mark in the top of the head.

The "V-shaped" nick in the skull of a modern infant baboon from the Gladysvale black eagle nest (left) mimics a similar mark on a fossil baboon skull from Taung (right).

Once again I returned to Johannesburg, my foot flat on the accelerator pedal of my old Toyota. This time I headed for the fossil lab. After laying my collection out on the lab table, I pulled out half a dozen of the fossils from Taung that were stored in the drawers. I needn't have even gone this far because the very first specimen was a small baboon skull with exactly the same depressed notch in the top of the head in exactly the same position. The V-shaped notches I had seen in other bones were matched by notches that were identical in the 2.5-million-year-old baboon skulls from Taung.

My heart thumped as my excitement grew. The evidence was perfect and it was in front of me. The Taung Child had been killed by a bird of prey.

I briskly made my way down the corridor to our resident Taung specialist, Jeff McKee, a young scientist who had come to South Africa in 1987 to take up a lectureship in anatomy after graduating with his Ph.D. from Washington University in St. Louis. I presented my case in detail to him, using the body-size graphs to show that Taung was an enigma and then telling him about the black eagle incident that had sparked the idea. I showed him the references on eagles and compared it to the species list at Taung, and finally, as the coup de grâce, I showed him the specimens from below the nest, their damage, and the similar damage patterns on the fossils from Taung.

"So what do you think?" I asked.

"I don't like it" was McKee's unexpected reply.

"But Jeff, the damage to the skulls?" I pleaded.

"Nope, that's not what collected Taung," he responded, gritting his teeth.

"Why not?" I asked, frustrated at what seemed like a brush-off.

"Because I know what collected Taung."

"What was it"

"I'll let you know when I finish the analysis," he remarked rather coolly.

It had been my intent to ask McKee to co-author the paper I'd planned to write on my bird-of-prey hypothesis because he had been working the site for years. But his rebuke had ended any possibility of that. I was still convinced I was on to something good, but McKee was adamant that I wasn't. I drew a deep breath and put what I thought was a gracious compromise on the table: "I'll tell you what, why don't we co-author a paper. I'll put in the case for a bird of prey, and you put the case for whatever your idea of the accumulator is. We'll leave it to the reader to decide."

"That's not the way science is done," he shot back. "You go ahead and write the paper and I'll write why you're wrong."

I left McKee's office demoralized, and returned to the fossil lab to lick my wounds and review the damage patterns again. While I was sitting at the large felt-covered table holding the fossils and modern bones, Ron Clarke, a research officer with the Paleoanthropological Research Unit, walked in. On the spot I made a decision.

"Can I show you something?" I asked.

"Sure," Clarke replied, "come into my office."

I put my case forward. When I finished, I just looked at Clarke across the desk.

"And?" I asked, with less confidence than I had before.

"My God, Lee, you've figured it out. It makes perfect sense!"

Clarke and I collaborated on an extensive paper that identified a bird of prey as having collected the Taung assemblage, including the Taung Child itself, based on damage marks preserved on the top of the head. Our major supporting criteria for the hypothesis were the damage marks on the skulls of the Taung hyraxes and baboons: double-depression fractures, V-shaped nick marks, and depressed flaps of bone that characterize raptor assemblages. In addition, the overall character of the assemblage did indeed match those collected by large birds of prey as I had first suspected with my crude body-weight assessments. We also settled on a most likely culprit species: the crowned eagle of Africa. This tremendous raptor is a specialist in primate hunting and has even been known to take human children. It can kill animals up to 70 pounds, and the variety of prey species we find at Taung fit in well with the crowned eagle's established behavior.

The bird-of-prey hypothesis was not just a good detective story.

It told us something more about the life of the australopithecines, poignantly revealing just how vulnerable these medium-size creatures were in their environment—preyed on not only by saber-toothed cats, hyenas, leopards, and the common carnivores, but even by ancient eagles. They were not the dominant animals in their environment, a far cry from what the later *Homo* species were like. Bob Brain was right. We were the hunted, not the hunters.

CHAPTER 7

LONG ARMS— SHORT LEGS

~

E ver since Raymond Dart's 1925 description of
Australopithecus africanus, this ape-man species has
been scientifically prodded and pulled, analyzed and
criticized, and alternatively reviled and revered by
those seeking to shape the human family tree. For the first 20
years following the discovery of the Taung Child—the Type, or
first specimen of *africanus*—there were few supporters for Dart's
claim that Taung was the intermediate form between ape and
human. The heavy pendulum of scientific approval, however,
swung in *africanus*'s favor during the late 1930s and 1940s as further finds confirmed that Taung was not an aberration. But by
the late 1970s, with the human family tree entangled by new
branches, *africanus* found itself out of fashion once again. In
particular, Donald Johanson and Tim White's incisiveness in cap-
turing the world's imagination with Lucy consigned *africanus* to
an evolutionary backwater as *afarensis* became the new bench-
mark in defining fledgling humanity.

But still, during all these years, first with Taung, then with
Mrs. Ples and the other three hundred or so known specimens of
this species that have since come to light, *africanus* was studied by
just about every single leading paleontologist from Berkeley

to Beijing. Yet a breathtakingly obvious and critical anatomical feature escaped them all. *Africanus* did *not* have the humanlike body proportions that everybody had assumed. It had long arms and short legs. How could this fact have been overlooked?

How did one of the world's most intensely studied hominid species fool our top scientists for 70 years? In early 1992 I found myself pacing around the university fossil vault pondering this problem as the newly discovered but fragmented *africanus* skeleton Stw 431 lay on its back on the green felt table in front of me. Seventy years since the Taung Child had first revealed an australopithecine to the world, new evidence from South Africa, in the form of this new skeleton, revealed an aspect of this species that we had never suspected. I was at once disturbed and excited by the problem that I had found within the Sterkfontein collection, a problem that seemed likely to force us to take a long hard look at our image of *Australopithecus africanus*.

"How could you have duped us for so long?" I wondered aloud at Stw 431.

I could see that Lucy was part of the problem. Because there was so much of her, and she, for the most part, was proportioned like a human, Lucy had become the model that governed our perception of all early hominid body shape. With relatively few postcranial bones from southern Africa—and these for the most part being fragmentary—and not one good partial skeleton until now, we as a scientific community had just assumed *africanus* would follow the same model as *afarensis*, namely having largely humanlike body proportions.

But the real reason, I believe, is that we got fixated and fooled by the skulls. The Stw 431 skeleton was the fossil that first alerted me to the fact that something was very wrong within the Sterkfontein hominid fossil collection. The fault lay actually with the way specimens were being interpreted. The discovery that *africanus* did not fit the mold of other hominid species when it came to body proportions had to do with a bit of luck and the quality of this one specimen. Stw 431 is probably the most complete *africanus* skeleton yet recovered. It was a fitting specimen to begin studying for my doctoral thesis on the shoulder girdle of the australopithecines. This partial skeleton found in 1989 by Alun Hughes had part of a right arm and shoulder girdle, a single rib fragment, most of the thoracic and lumbar vertebrae, and a sacrum and pelvis. One solitary tooth was all that was left of the head. Despite its incompleteness, Stw 431 was immensely valuable. It was the first time that we had ever had a southern African specimen with remnants of both the upper limb and lower limb preserved.

I was particularly interested in working with this specimen for my Ph.D. for two reasons. Stw 431 had a scapula and clavicle, and to date relatively little work had been done on the new post-cranial material from South Africa. Also, the specimen would shed some light on the raging debate about whether these ape-men—and australopithecines in general—were fundamentally arboreal or terrestrial. I had specifically chosen the shoulder girdle to study because any climbing that these early hominids did would require significant involvement of the upper limb. The newly discovered material from Sterkfontein had within it a decent-size collection of shoulder girdle elements, an area that had not been studied extensively in any early hominid species due to a lack of the relevant remains. So in beginning my doctoral studies, I was convinced that a comparative study of the shoulder girdle elements of *africanus* could bring about some resolution to this ongoing controversy in the field.

In the eighties and early nineties, the arboreal-terrestrial debate centered around *afarensis*, and in particular just how human it was from the neck downward. This debate was surprisingly vociferous, so much so that "camps" developed among scientists with "arborealists" on one side and "terrestrialists" on the other. Arborealists largely viewed *afarensis* as a tree climber, more comfortable hanging out in a forest canopy than walking as a human on the ground. Most European scientists at the time were arborealists—the French being the most ardent—while in the United States, their sympathizers were to be found mostly at the State University of New York at Stony Brook. The arborealists argued that in cases where *afarensis* did walk on two legs, it did so only in a marginally humanlike way.

For the most part, *afarensis* arborealists were skeletal "splitters." They viewed a specimen more from its specific features than its general morphology, placing great weight on individual and linked features in the skeleton that would indicate climbing adaptations. It was pointed out, for instance, that *afarensis*'s fingers and foot bones were longer and more curved than in modern humans, suggesting these appendages were designed more for tree climbing. Peter Schmid of the University of Zurich demonstrated convincingly that Lucy's thorax was funnel-shaped, which was more ape-like than the barrel-shaped chest of humans. Her ilia, the blade-like structures in the pelvis, were wide and flaring, unlike the more curved ilia of humans. The French scientific duo of Brigitte Senut and Michele Tardieu argued that based on specific characters found in the proximal tibia, *afarensis*'s knee joints had greater mobility than in modern humans, which would have been an advantage in climbing trees. The case was also made by Bill Jungers of Stony Brook that, at least in Lucy, the femurs were somewhat shortened in relation to the upper-limb length, giving her a slightly "long armed" appearance. This combined package of features, the arborealists would argue, added up to an animal in transition. There was no doubt that *afarensis* was bipedal on the ground, they conceded, but it clearly had retained features that showed it was both dependent on the trees and a capable climber.

The terrestrialists saw *afarensis* as more comfortable walking around on two legs on the ground than living in trees. They were essentially skeletal lumpers, those who took as their departure point the fact that Lucy was bipedal and that all other features were secondary. They had as their senior champion the American scientist Owen Lovejoy, who was originally involved in the description of the Hadar material. The terrestrialists concentrated more on the bigger picture of *afarensis*'s skeleton than focusing on the smaller details. They would point to the obvious, that Lucy had largely humanlike limb proportions, namely long legs and short arms, compared to all other apes, and that this was clearly like the condition found in modern humans. Since Lucy had this form, then *afarensis* had this form. And since *afarensis* represented the mother species of all later hominids, then humanlike terrestrial dependency had appeared very early in the hominid lineage.

"Could they climb trees at all?" I once asked Lovejoy.

His reply was: "Once they had shifted that whole morphological package toward terrestrial bipedalism, they were committed to the ground."

The intensity that accompanies any discussion about human origins was not absent from this tree-climbing debate. In fact, in the late eighties and early nineties there was a period of intense disagreement between the camps, with sensitivities such that the interpretation of a single muscle's function could lead to public character assassination by scientists. At conferences, scientists could find themselves cast out from their groupings for merely speaking to a member of the opposite camp. Students, like myself, found ourselves in the middle of a cold war between senior scientists questioning what was essentially untestable without a time machine. But at the time, I was a committed arborealist. Why would *afarensis* have retained any of these primitive climbing features if it wasn't using them? I would later change my mind.

So where did *africanus* fit into this picture? *Africanus* was viewed by mainstream paleoanthropologists as a species descendant from *afarensis* that probably had little to do with the evolution of the genus *Homo*. Using *afarensis* as a comparison, *africanus* was seen to possess an abundance of derived traits, characteristics that seem to have developed away from a primitive state. For example, *africanus* molars and premolars were relatively large and thus perceived to be derived from the primitive small-toothed condition found in *afarensis*. This was a point established by Johanson and White in the late 1970s to prove that *afarensis* was in fact the mother species of *africanus*. They showed that *africanus*'s premolars were becoming "molarized,"and the size of the canines in relation to the other teeth were reduced—both of which to them indicated a move away from the primitive small-toothed big-canine condition found in *afarensis*. In addition, they argued, cranial capacity increased slightly in *africanus*, which is what one would expect in the evolution of a later hominid.

Most researchers, and most textbooks at the time, adopted the view in which the *afarensis* stem gave rise to two or even three forks. One of the splits led to *africanus*, which, because of the molarization of the premolars and the general increase in dental size, seemed to be the best candidate ancestor for the lineage of robust ape-men, including the southern African variant *Australopithecus (Paranthropus) robustus* and the East African version *Australopithecus (Paranthropus) boisei*. The other fork passed through a large question mark between 2.8 and 2.0 million years until it reached out to *Homo habilis*, which was the direct ancestor of *Homo erectus*, a species that would eventually evolve into *Homo sapiens*. Some more progressive phylogenies, particularly those constructed in the late eighties and early nineties, would take into account the increasing diversity in the early Pleistocene and would include *Homo rudolfensis*, a little known species that existed at roughly the same time as *habilis*, as well as *Homo ergaster*, a species name usually considered synonymous with "early African" *Homo erectus*.

But even by the late eighties, problems with this relatively simple evolutionary tree were already emerging. These difficulties came in the form of the discoveries of two skeletons attributed to *Homo habilis*, one from Olduvai Gorge labeled OH 62, discovered appropriately by Tim White and Don Johanson, and the other from Koobi Fora, labeled KNM-ER 3735. These two finds, while undeniably fragmentary, both were independently described as possessing a surprising body shape. Each skeleton seemed to have had long arms and short legs. This raised a question mark in some scientists' minds as to whether *afarensis*, which had more human-like limb proportions, could have given rise to a *Homo* species that in some respects appeared more primitive. Despite the complications that were beginning to emerge through the likes of OH 62 and KNM-ER 3735, by the end of the 1980s, the human family tree still looked fairly straightforward: *Australopithecus afarensis* remained firmly at the stem as the mother species of all subsequent hominid species. Nothing more primitive and, maybe more importantly, nothing older had been found. A great point in Lucy's favor was that she was supported by secure dates. By the mid-eighties, geologists working in East Africa had perfected the use of potassium–argon dating, and the dates obtained on volcanic ashes firmly placed *afarensis* in a temporal range between 3.7 and 2.8 million years ago. This made it the oldest definitive hominid and cemented the proof that our earliest ancestors came not only from Africa, but specifically from East Africa.

The fossil evidence suggested that our earliest forebears were up on two legs before their brain size increased dramatically and before major restructuring of the face and dentition occurred. The *afarensis* fossils from Ethiopia and Tanzania indicated that humanlike bipedalism had arisen early on our family tree and that for the next couple of million years, the most interesting things happening from an evolutionary perspective were going to be found in the heads.

When I began my Ph.D. research on the shoulder girdle, I was convinced that *africanus* played some middle-of-the-road role in human evolution. Its well-studied heads fitted the pattern of a daughter species of *afarensis*, and there was little in the literature to suggest that its body would not follow the largely humanlike pattern set by *afarensis*. I began my analysis of the Stw 431 skeleton nonchalantly enough, not really looking for anything out of the ordinary. But when I began the rather tedious process of measuring and describing the shoulder girdle elements from the collection, I was not prepared for what I would find.

I started my study with the examination of the scapula fragment from Stw 431. My first impression was that it was very large. Taking various measurements of the specimen, I was surprised to find that if I compared my measurements of Stw 431's scapula with those of a modern human sample, that person would be as large as a modern muscular football player. This revelation stopped me in my tracks. This didn't make any sense. *Africanus* was meant to have been a chimpanzee-size creature, not some lumbering giant. I decided immediately to abandon the sole study of the scapula and to measure the other parts of the skeleton, applying the same comparison with human elements. By comparing Stw 431's distal humerus, ulna, and radius, I quickly found that they presented the same pattern: They were all in the upper range of human variation in overall size. But I didn't expect, when I took measurements of the pelvis and sacrum, to see that these bones fell into the lower-size ranges of human variation. These results were unexpected and puzzling. Something was wrong with Stw 431's body proportions; it seemed to have large upper limbs and a tiny pelvic girdle.

In order to get a clearer picture of the problem, I went to our cast collection and pulled out casts of the upper limb material from Hadar. This collection was largely made up from *afarensis* individuals known as the First Family, found by Johanson in Ethiopia during the field season following Lucy's discovery. One could see as soon as they were laid out on the table that some of these upper-limb fossils were as large as those of Stw 431, but some were much smaller. I then retrieved the cast we had of Lucy and laid the corresponding elements of her skeleton out on the green cloth alongside the remains of Stw 431. In general, Lucy looked tiny by comparison. Her upper limbs were just over half the size of the corresponding parts of Stw 431.

But the pelvis was another story altogether. Oddly enough, the Stw 431 pelvis didn't look much bigger than Lucy's. If the upper limbs were anything to go by, Stw 431 was much bigger than Lucy. There should have been a far greater size difference. Between the two, Stw 431 should have possessed an ilium at least twice the size of Lucy's. I held the ilia of Stw 431 in my left hand and Lucy's in my right. I was astounded to see they were about the same height. This just didn't make sense. I put the bones down and compared the sacra, or tailbones, of the two hominids. While Stw's 431 sacrum was slightly more robust in construction than Lucy's, the latter was actually wider. What the heck was going on? It all seemed so disproportionate.

The 431 skeleton first led the author to suspect that
A. africanus *had different body proportions from the Lucy species* A. afarensis.

I needed a broader visual context to examine this strange phenomenon. I carefully reassembled the specimens I'd been studying to make up the entire skeletons of Stw 431 and Lucy and laid them side by side on the green cloth next to a human skeleton. The picture immediately became clearer. As I expected, both upper and lower limbs of Lucy looked equally proportioned, and not far off the anatomical relationship of the human bones. But Stw 431 looked absurd. It was completely top heavy, with these great big upper-body parts and this petite lower-body structure.

Even then, when things should have been glaringly obvious, I remained confused, stuck in my perception that Lucy's body shape represented the more primitive form and that all subsequent hominids should look more humanlike in their build. Certainly that was the case with the skulls. There is no denying that Lucy's cranial features looked more primitive than that of *africanus*. Perhaps Stw 431 was a pathological abnormality. I needed more *africanus* bones. Over the next few hours, I pulled all the *africanus* postcranial bones out of the safe and scrutinized them with suspicion. They told the same story. Every upper-limb specimen I had for *africanus* was big and every lower-limb element was small, precisely reflecting the pattern that I'd picked up in Stw 431. This is crazy, I thought to myself at the time. There's something very wrong with the whole collection. I tried to rationalize the situation. Could it have been a gender difference creating this odd sample? Maybe the accumulating agent at Sterkfontein had somehow collected just the upper bodies of males and lower bodies of females. But not only did this seem absurdly improbable, Stw 431's proportions said that this was the *africanus* norm.

I then took a chimpanzee skeleton from the drawer and laid it on the table next to the trio of skeletons already there. The logic of it all then suddenly dawned on me as if someone had splashed cold water on my face. My preconceptions were pierced in an instant as I realized my eyes were not deceiving me. *Africanus* had a massively built upper body with long arms, and yet its lower limbs resembled in relative size and length those of a chimpanzee. A different kind of biped than Lucy. I was thunderstruck. This observation flew in the face of everything that I had been taught, not only about the species but about the whole story of early human evolution. Long arms and short legs! I poured myself a strong cup of coffee and sat down to distill the implications of this morphological revelation.

There was no denying that *africanus*'s skeleton in its proportions looked far more primitive than that of *afarensis*. Yet *africanus*'s skull was more "modern" than that of *afarensis*. In conventional evolutionary terms this was nonsensical. How could a species of hominid advance from the neck upward while the rest of the body regressed into a more ape-like form? I could reach only one logical answer. *Afarensis* could not be the mother species of *africanus*. They must have a common ancestor—one that had yet to be discovered—that had *afarensis'* skull, or something like it, and something akin to *africanus*'s body, in which case they would only be sister species. In any event, one thing was for certain: What I was seeing before me on the green felt table in the hominid vault at Wits meant that the whole way in which we drew the human family tree might be wrong.

How could such a difference between *afarensis* and *africanus* have gone unnoticed? Perhaps it wasn't actually all that surprising. The postcranial record of *africanus* has been relatively poor to begin with and not much of it has been published. Additionally, *africanus* as a species so far appears to have been endemic to South Africa, a country not only distant from the great fossil fields of East Africa but one that has languished in a scientific wilderness for decades. In fact most of the published South African material on *africanus*'s postcranial anatomy has been based on the fossils recovered by Robert Broom from Sterkfontein in the 1930s and 1940s. These fossils consisted of a few dozen samples and one crushed vertebral column, pelvis, and associated femur (Sts 14). There was also the small amount of material collected by Kitching and Dart at Makapansgat in the 1940s—two ilia, the pelvic blades from the pelvises of juvenile hominids. About all that scientists could conclude from these fragmentary fossils was that *africanus* was not nearly as sexually dimorphic as *afarensis*.

The difference in size between male and female specimens in the South African sample as a whole appeared to be closer to that seen in chimpanzees, while *afarensis'* sexual dimorphism was closer to that seen in gorillas. The bits and pieces of the South African sample that were better preserved, a scapula and crushed humerus (labeled Sts 7) and some hand and foot bones, were enigmatic. Granted that both Robert Broom and John Robinson had early on described at least the scapula fragment of *africanus* as being orangutanlike, but it was so fragmentary. Furthermore, there was no well-preserved scapula from the *afarensis* collection to compare to this one enigmatic bone, so this nonhumanlike pattern was largely overlooked. Researchers tended to lump the small South African sample into the larger and better-preserved *afarensis* continuum. And what difference would it make anyway: There has been little contention that humanlike bidepalism emerged early in human evolution.

I believe that the main reason that no scientist had yet picked up this critical difference between *afarensis* and *africanus* was that we had all been blinded by the skulls. The heads of these early hominids seemed to track time so well that the rest of the anatomy was assumed to have followed suit. Given that it was clear that both species walked on two legs and had roughly the same kind of pelvic structure, this was not a bad assumption. I knew from the moment that the differences between *afarensis* and *africanus* emerged that I had my work cut out for me. What was really just a hunch had to be backed up with a watertight case. I set to work reexamining the many postcranial bones in the safe beside the Stw 431 skeleton and, with disappointment, noted that there was a serious lack of complete specimens. But this scrappy sample was probably the best sample of early hominid postcranial remains in the world, so I knew that I had better figure out a way to work with them.

The first step was to perform simple linear comparisons between the *afarensis* and *africanus* samples. I went into the department's large collection of human skeletons and took measurements of the upper limbs, sacrum, and pelvis areas that were preserved in Stw 431. Additionally, I took measurements of the acetabulum—the pelvic cavity in which the femur rests—to get an idea of how big the femur of Stw 431 might have been. By dividing any given lower-limb measurement by an upper-limb measurement—say, the medio-lateral diameter of the sacrum by the medio-lateral diameter of the distal humerus—I could create a percentage, or index, that represented the relative size difference between these joints in a human or a chimpanzee. If you do this at two joints or morphological areas in enough humans or apes, add them together and divide by the number of individuals measured, you can create a mean and range for a sample. Thus at any two joints of the upper and lower limb, I could get a rough idea of the size difference between the joints.

In this way, I could do a simple test as to whether my visual concept was correct that the upper limbs of Stw 431 were relatively much too large for the pelvis and in differing proportions to Lucy and to a human sample. With just a small sample of humans, I quickly confirmed that Lucy's upper- and lower-limb joints at least were broadly comparable to the same indices I found in humans. Stw 431, I noted with satisfaction, was completely different. Continuously, across several joint areas, I found that the indices reflected an upper limb that was large relative to the pelvic girdle. The closest match I could find for most indices was when I compared Stw 431 with a chimp sample. But I had a thesis to complete, and so I took up the hobby of adding more and more data to my sample as the months went by. Eventually, my database was large enough that I was convinced that *africanus* was proportioned more like an ape. Furthermore, I had come to realize that it was not just a matter of mass differences, but the actual relative limb lengths of *africanus* were more like those of a chimpanzee than those of a human. *Africanus* was built pretty much like a bipedal nonhuman ape.

After over a year and a half of part-time work, and even though the picture being painted by the indices looked good, I knew that it would take more than the simple creation of ratios to convince my colleagues around the world that they'd all been wrong in their understanding of early human evolution. There were too many vested interests in the science and too many reputations at stake for some upstart from South Africa to turn all their thinking upside down. Yet the evidence in my own mind was stark. Either the thinking that *afarensis* had given rise to *africanus* was plain and simply wrong—no matter how neatly the heads seemed to develop—or there had been some extraordinary reversal in human evolution that saw a more advanced species reverting to primitive body proportions.

I finished writing up the results and submitted my Ph.D. during the course of 1994, freeing up more time to concentrate on the long arms—short legs problem. One of the first things I realized was that I needed help in preparing my case. I neither had the expertise nor the modern sample of apes and humans to do the kind of statistical analyses that would prove this point to the scientific establishment. I needed to find the paleontological equivalent of the best attorney available. The only person that I could think of with such a database, and with the expertise to boot, was Henry McHenry of the University of California, Davis, one of the most respected scientists in this field, a specialist on body proportions.

For the past two decades, McHenry had centered his work around the prediction of limb lengths and body weights in early hominids. During this time, he amassed a data bank of hundreds of humans and apes for which he had body weights, joint sizes, and bone lengths. He was also a committed terrestrialist. If there was one ally I needed to muster in a field of skeptics, it had to be McHenry. Access to this information was crucial for what I wanted to do with the *africanus* sample because our specimens were by and large very fragmentary. McHenry had the key to proving that something different was going on in *africanus*. I just had to persuade him to let me use that key.

It happened that in April 1995 both the Paleoanthropology Society's and the American Association of Physical Anthropologist's annual meetings had been scheduled to be held in Oakland, California, not far from McHenry's base in Davis. I knew there would be a good chance that he would attend, but just to make sure, I wrote him a short note asking him to make some time for me at the meetings to discuss this matter.

The Oakland conferences are memorable for a particularly vitriolic session of the Paleoanthropology Society Meeting that degenerated into a screaming match over the ownership of Ethiopian fossils. A Rutgers University fieldworker accused Don Johanson's Institute of Human Origins of appropriating a fossil hominid from their search area during the previous Ethiopian field season. This accusation prompted a sharp exchange between a number of well-known scientists, and even the brave few who tried to cast themselves as peacemakers found themselves wrapped up in the whirlwind of controversy. The conference conveners hastily announced a short break to allow tempers to cool, and it was during this intermission, amid a cluster of delegates muttering darkly, that I met McHenry to ask for his assistance. Equipped with a handful of carefully selected casts representing the upper limbs and pelvis of Stw 431, I made my case. First I showed McHenry the bones. I pulled the pelvis from my conference bag followed by each limb bone that I'd hoarded in my pockets for the occasion. I lined up the casts on the carpet of the hotel lobby to illustrate the critical differences in size that would explain my obsession with why long arms and short legs were the essence to understanding early human evolution. McHenry agreed that these bones looked a bit out of proportion, but he wasn't entirely convinced. I then pulled out my data sheets from the sample of human and ape skeletons I had measured back in Johannesburg and proceeded to sketch out my hypothesis. Finding myself with no paper and without a table to work on, we crouched on the hotel lobby floor and I used a paper napkin as my sketchpad. I carefully outlined my idea, scribbling with a ballpoint pen on the frustratingly soft napkin surface. If one divides any major measurement of the pelvis by any major measurement of the upper limbs, one can arrive at a simple ratio. If I applied this to Stw 431, I got an unusually low index, more like that seen in an ape. But in Lucy and the humans I had measured, I was getting nearly a one-to-one ratio.

McHenry played the role of devil's advocate throughout, pointing out to me that simple linear measurements could be misleading and that I certainly couldn't use just one individual to represent a whole population. I was prepared for this. I showed him the index range for the whole sample from Sterkfontein; how the upper limbs were all big, while the lower limb material was all small. By this time, we had attracted a small audience. South African fossils were virtually unknown to the outside world in 1995 and people were intrigued. Other scientists began to add their input and ideas. Maybe it was a taphonomic problem, suggested one. Possibly some predator had selected male upper limbs and female lower limbs, added another rather fancifully. I had thought these problems through before and was ready with an answer.

"If we had only been dealing with a small sample, say two or three individuals, I might have bought this line of reasoning," I replied. "But what do you say to a sample of over 70?"

Silence fell around me. Finally, McHenry turned toward me, picked up the last napkin I had been drawing on, folded it with thoughtful deliberation, and tucked it into his jacket pocket.

"Ok," he said, "I get it, but I'm going to have to come and see this for myself."

I must have grinned from ear to ear because he laughed.

"How about later this year?"

McHenry arrived in Johannesburg in June 1995 and committed himself to a one-week intensive study of the Sterkfontein collection to see whether he would arrive at the same conclusions I had. He based himself in the lab at Wits and got down to work almost immediately. The first thing we did was to sort the collection into usable and unusable components. McHenry had a rule for his analysis: No children allowed. He only wanted the bones of adult specimens, and these had to be either fairly complete joint surfaces or whole bones. Juveniles would bias the sample because their size would depend upon their age at death. So we separated every juvenile specimen from the collection. We then pulled out any postcranial bones that did not have enough articular end or shaft preserved to measure. This still left us with a relatively large sample: 48 elements in total. At the time it represented the single largest assemblage of adult early hominid postcranial elements available from one site for any single species of early hominid. Once we sorted this collection, we began measuring the individual pieces.

This was a vital undertaking upon which my whole theory would rest. Every measurement has precise anatomical points that the end of the calipers must touch. Due to variation, these points can be hard to identify on modern bones. On fossils that are weathered and damaged and sometimes broken, the identification of an anatomical point can be extremely difficult. McHenry and I would sit and discuss these difficult decisions, trying to come to an agreement on where a particular point might be; a small discrepancy in position of the caliper tips might translate into a large error. Additionally, each measurement had to be taken at least three times to ensure an accurate reading, and we would often pass a specimen back and forth to see if we came up with the same result independently. McHenry would cheerfully try and find small measurements in the upper limb and large ones in the lower limb, but the pattern of big upper limbs and small lower limbs continued. After almost a week, we had all the measurements.

Throughout the process, McHenry noted the evolving trend and admitted that he was becoming convinced that my theory may have the backing of the physical evidence. On his second to last day in the lab, he suggested that we pull all of the material from South and East Africa out of the safe and lay it out on the big felt-covered worktable in the middle of the lab—to get a "visual" on our results. Cramming almost 100 postcranial fossils of early hominids on that table was thrilling—I hadn't actually done this since my initial revelation in 1992. We put all the limb elements into groups of like elements, upper limbs at one end of the table, thorax in the middle, and lower limbs at the other end. We then laid the cast of the Lucy skeleton and some of the other *afarensis* material beside the South African fossils.

The bones spoke for themselves, giving the measurement results visual support. There were no big lower-limb elements and no small upper-limb elements in the Sterkfontein collection.

"We'll have to see what the regressions show, but I think it looks good," McHenry said as we surveyed the collection of fossils and casts. "I'll be able to have the body-weight predictions and limb lengths for you within a couple of weeks, and we'll see what those say, but I'm pretty convinced. There's something different going on here."

McHenry seemed personally convinced by the data, and we discussed the implications of "redesigning" *africanus* on our understanding of human evolution. I was thrilled that the world's leading fossil-bone analyst agreed that I'd found something previously overlooked, but I was completely surprised that his conclusions were so different from mine

McHenry argued that evolution did not necessarily progress in a linear format, and that we should not get caught into believing that "primitive" necessarily meant older, and that "derived" suggested advanced. Evolution, he said, narrowing his eyes to underline the gravity of his statement, had temporarily gone backward. What on Earth did he mean? Evolution is a response to a changing environment, he continued, which meant that animals adapted their morphology to cope with shifting habitats. We therefore should not assume that evolution meant a straightforward advance from our idea of an ancient ape to the modern human form with no deviations along the way. The skulls, I thought. McHenry is of the school that believes the heads led the way. I voiced my skepticism about being fixated by skulls. But, he asked, did I not agree that the skull of *africanus* carried features that suggested it was a descendant of an *afarensis*-like ancestor? Yes, I conceded, but quickly pointed out that the postcranial features that he had studied earlier in the day told the opposite story. McHenry would not be thrown off his stride. You have to admit, he said, *habilis* looks like it descended from *africanus* and not *afarensis*? Yes, I responded cautiously.

"Well then," he continued with a final flourish. "It makes sense to me that the southern African ape-men must have undergone some sort of reversal that forced a 're-evolution' of long arms and short legs. Maybe they were forced back into a more wooded environment in this region and therefore took to the trees again, so to speak." If this was the case, he concluded, then the evolution of long arms and short legs would have been an adaptation to a changing environment and would simply mimic the presumed "primitive" condition.

"I can't agree," I said. "It makes more sense to me that *afarensis* and *africanus* were simply sister species. I think they shared a common ancestor that would have had the primitive head of *afarensis* and the primitive long-armed and short-legged body of *africanus*."

He looked interested but unconvinced.

"You know what, Henry," I continued, "The simplest answer may be that we have just not found the missing link between the two forms. Who knows what still might be found out there?" I waved my hand in a general northerly direction as McHenry shook his head vigorously. He conceded that the fossil record was by no means complete and that a new "missing link" might emerge from some African sediment, but he stuck to his view that *africanus* was displaying a simple morphological reversal. We continued the discussion, but no amount of intellectual parrying and thrusting on either of our parts led to consensus.

"Ok, so we're seeing long arms and short legs," said McHenry, leaning back in his chair with a triumphant smirk flitting briefly across his face, obviously determined to have the last word. "But what if there's more than one species in the collection? What if there was more than one kind of hominid in the sample, and we've been assuming that we're dealing only with a single species? That would skew our results."

I'd been avoiding this question for the past two weeks. But now McHenry had hit on one of the biggest problems facing the Sterkfontein sample. Not only were there the problems in the assemblage of provenience and mixing to deal with, but for years researchers working on the cranio-dental material, a much larger sample, had been suggesting that there might be at least two species of australopithecine in the gracile material from Member 4.

The Sterkfontein "two species" question has been, and continues to be, the subject of a fairly vigorous debate at Wits. Given the jumbled taphonomy of the South African fossil sites, and the degree of variation that *africanus* displays at the best of times, it will be a difficult argument to resolve. The first salvos in this debate had been fired by Ron Clarke in the early 1990s, when he suggested that Sterkfontein Member 4 samples represented the beginning of a speciation event—that is, the emergence of one species from another. Clarke suggested that the anatomical range of *africanus* material from Sterkfontein was too great to be ascribed to either a single species or to males and females within that species. He'd based his argument largely on one specimen, Stw 252, which he said appeared to be more derived than most of the ape-men fossils excavated from the site.

Tobias had rejected Clarke's argument, pointing out—with some justification—that Clarke had chosen a particularly damaged specimen in Stw 252 upon which to make such a sweeping statement. Tobias also pointed out that Clarke had only used a few morphological traits to make his case and had no statistics to back his argument. Clarke conceded that there was a level of subjective interpretation in reconstructing the skull, but insisted that the teeth, which were in relatively good condition, were far larger than the accepted norm in the gracile ape-men. This, he said, couldn't be argued away on the basis of damaged skulls.

I looked at McHenry. "Well, okay, let's say that's the case. Say there are two species mixed up in the Sterkfontein assemblage and not one, then surely both species must have the same or similar body proportions, long arms and short legs. Otherwise, we'd see more overlap in the sample. We'd get a mixed signal." I thought about the problem a moment more. "Secondly, I've gone through the whole collection of postcranial bones while working on the descriptions. There is a bit of variation in morphology, but not what one would expect from two different species. At least not two different species with different body proportions."

"Ok, I agree with that, but we should look at our overlapping specimens closely to see if anything shows up." McHenry pulled out the list of specimens and their corresponding measurements. "These are the ones," he said, reading off the numbers. I wrote down the specimen numbers that represented the largest lower-limb elements and smallest upper-limb elements from our sample.

"I'll check these in the morning before you leave," I said.

The next morning I looked at the eight largest lower-limb and smallest upper-limb specimens. In particular I examined their provenience and morphology. Three specimens showed some unusual features. The smallest upper-limb element was a proximal humerus, and just by coincidence, or maybe not, it came from the same grid square as one of the largest lower-limb elements, a proximal tibial specimen. The third specimen was a fairly complete femur, Stw 99, which was listed in the catalog in Alun Hughes's handwriting as "Member 4 or 5 *Australopithecus robustus*." I had checked on this specimen before. I had even gone to the site and found the position of its recovery, and Clarke had assured me that this specimen was almost certainly from Member 4. I reported my results to McHenry. "Ok, we'll keep those specimens in mind as we go, but I'm happy to go with the idea that if there's more than one species, we're either not sampling it in the adult postcranial material or its got nearly the same body shape."

As promised, within two weeks of his return to California, McHenry faxed me the results of his body-weight prediction for the upper- and lower-limb fossils. The results confirmed what we had suspected. When using a regression that assumed the sample of the upper-limb specimens was human, there were only four that gave a body weight below 108 pounds—and the smallest of these were two that gave a predicted weight of exactly 92 pounds and one that gave an estimate of 86 pounds. The largest estimate was 187 pounds. The mean estimated body weight for the whole upper-limb sample came to 112 pounds, meaning that if you had pulled an "average" *africanus* upper body from a sample, it would weigh around 110 pounds.

As we had also predicted, the lower-limb elements gave completely different results in that there were only four lower-limb specimens that gave body-weight predictions over 88 pounds, and only one of these was above 99 pounds, at 108 pounds. The mean body weight for the lower-limb elements was a paltry 75 pounds. We even had one adult individual with a body-weight estimate in the lower limb of a paltry 22 pounds! It was as if we had sampled two completely different species in the upper and lower limbs.

The human regression used to predict body weight seemed to be overestimating the body weights of the upper-limb material and underestimating the weight of the lower-limb material in *africanus*. If this had been a random sampling of humans, one would expect to see both big lower limbs and small lower limbs as well as big upper limbs and small upper limbs. In other words, if *africanus* had been proportioned like a human, we should have either got mean upper-body weights of 75 pounds to match to lower body weights, or we should have found that the mean of the lower body-weight prediction was around 50 pounds.

Most poignant were the results for the Stw 431 skeleton. Its upper-limb mass predictions were approximately a whopping 125 pounds. The lower-limb predictions made from the joint sizes at the acetabulum, sacrum, and pelvis averaged only 65 pounds, nearly the mean of the sample and totally different from what one would expect in a human.

When we ran the same human regressions on the *afarensis* material, we not surprisingly came up with very different results. In the Hadar sample, the numbers scattered widely. There were big and small upper-limb elements and big and small lower-limb elements. When the Lucy skeleton was measured and her weight estimated using the human regression, she came out with a lower-limb estimate of about 53 pounds and an upper limb estimate of around 57 pounds. Just what one would expect from an animal proportioned very much like a human.

When we ran the *africanus* sample using a chimpanzee regression for body-weight predictions, we ended up with a good scatter of big and small lower-limb elements and big and small upper-limb elements, just as you would expect in a normal chimpanzee population. Using the chimp data, the Stw 431 skeleton even came out with a lower limb prediction of about 90 pounds and an upper-limb prediction of 101 pounds. So the chimpanzee body shape was a better predictor for mass in *africanus*. When we ran the same regression on *afarensis*, the numbers were flipped, so that the lower limbs were estimated as way too large, while the upper-limb weights were underestimated (chimps have relatively much bigger upper limbs and much smaller lower limbs than humans). Limb lengths came out the same way, with *africanus* showing much more ape-like proportions than *afarensis*.

This was the proof I needed.

CHAPTER 8

KNEE-JERK
REACTION

～

The peri-urban sprawl of Nairobi came into sight over dusty acacia trees dotting the yellow-brown grassland as my South African Airways flight approached Uhuru airport. Flying is conducive to the freewheeling flow of thought that connects disparate ideas into a meaningful continuum, and *africanus* seemed to have staked some claim of ownership over my mental facilities. I'd spent the four-hour flight from Johannesburg to Nairobi picking through the implications of the long arms–short legs theory. It was late 1995, and I was starting to feel that I was gaining headway in making sense of this increasingly mysterious ape-man, but there was still something nagging me about the Sterkfontein collection.

I stared out the window at the East African landscape and reflected on how good the past 12 months had been for me—and *africanus*. My challenge of *afarensis* as the mother species had been taken seriously in the forums in which it had been raised, but I was a long way off from having the idea universally accepted. Just how difficult shaking up the human family tree would be was brought home to me when I ran into opposition from the very quarters where I was expecting the most support— from my colleagues at Wits. In particular Phillip Tobias did not like the conclusions I was reaching, even though he had long argued against *afarensis* and was a traditional *africanus* supporter. As the aircraft banked and prepared for landing, Tobias's warning rang through my mind, as it had countless times since he'd first uttered it.

"If you keep going in this direction, you're going to turn *Australopithecus africanus* into an ape."

I still had not got over the shock of his words. At the time I'd looked in vain for the familiar wry flash of humor in his eyes, but the gaze that had greeted me across the fossil bones in the hominid vault at the University of the Witwatersrand was unambiguously one of disapproval.

Far from wanting to relegate *africanus* to the status of marginal deadwood in the human family tree, I was intent on putting this extinct little ape-man back at the base of the evolutionary trunk. Yet the professor, whose life work had gone into championing the protohumanness of *africanus*, was clearly concerned that I was playing into the hands of his intellectual enemies. I was thrown by his reaction, partly because he was my mentor, and partly because I had expected his support. He had listened to me for an hour as I enthusiastically outlined how Henry McHenry and I had found that *africanus* had this massive upper body atop these relatively petit lower limbs. Tobias had listened closely as I concluded that the sum of these body parts made it improbable that *africanus* was a descendant of *afarensis*, and that the Sterkfontein speciation event suggested that early *Homo* seemed to have been derived from late *africanus*. But at the end of my explanation he shook his head, and the dismissive tone of his voice indicated that although he agreed with where I wanted to place *africanus* in the human family tree, he remained unconvinced by my reasoning. I masked my disappointment with silence. Clearly he was concerned that his baby was about to be thrown out with the bathwater; that *africanus* would be a victim of its own morphology and would be relegated to ape status, and that *afarensis* would remain on the protohuman pedestal. If Tobias would not support me, how much chance would I have with scientists like White and Johanson, whose reputations were so intrinsically vested in *afarensis*?

I came down to Earth with a bump as the aircraft undercarriage bounced on the Nairobi tarmac. The plane taxied down the runway, past the stainless-steel fuel tanks gleaming in the afternoon sun, and came to a halt outside the modest international arrivals terminal. I remembered how strange the light and the air had felt when I first arrived in Kenya six years previously, and how alien East Africa had felt. Landing at Uhuru airport now was like coming back to a favorite place. Even the machine-gun-armed military guards that stood around the airport concourse and in customs seemed more familiar and less threatening than they did when I had first begun my African journey.

There's a strange gap that exists between East Africa and South Africa in relation to the search for human origins. The two regions are undoubtedly the most important outdoor laboratories in the world in providing us with original material about our ancestry. And while a strong thread of individual scientific cooperation has stretched back over half a century, there has also been a degree of competitiveness over which area is the most paleontologically significant. I consciously warned myself not to get jingoistic in the company of my East African colleagues, and to play down my feelings that, paleontologically speaking, South Africa would be where it was all happening in the late nineties. Besides, it was only through a process of cross-referencing material from the different regions that the science as a whole could move forward, and I was in Kenya to compare the body proportion work that I'd been doing on *africanus* with original material from East Africa. In particular I was interested in seeing what light the Turkana Boy and KMN-ER 3735 (the fragmented *Homo habilis* skeleton that reportedly had long arms and short legs) could throw on the long arms–short legs theory for *africanus*.

After clearing customs, I stepped out into the arrival hall and was met by a smiling Meave Leakey. Meave is a zoologist and paleoanthropologist with the Kenya National Museum. She also happens to be Richard Leakey's wife—although don't ever make the mistake of defining her in such a subordinate way, as her contribution to the science has been enormous. She's not a hard person to recognize in a crowd. Slender with graying hair, she displays a quiet, gaunt determination that epitomizes the spirit of the African explorer.

"How was your flight up?" she asked as we walked out into the hot sun toward her Land Rover.

"Super," I replied. "I just keep forgetting how far South Africa is from Kenya." She nodded her head in agreement. She knew I wasn't just talking about the mileage.

I was excited about meeting Meave. She's an engaging colleague at the best of times, but on this visit we really had something to talk about—*Australopithecus anamensis*. This was a new species of ape-man that her museum research team had found at a site on East Lake Turkana just a few months before. With an estimated age of 4.1 million years before present, *anamensis* —*anam* is the Turkana word for "lake," hence "southern ape of the lake"—is one of the earliest hominids discovered so far. It is also one of the few fossil fragments to have emerged from beyond the four-million-year-old barrier, and is therefore tantalizingly close to the five-million-year-old milestone, which is when we assume we diverged from the lineage of ancestral apes. My intense interest in *anamensis* was not just in its antiquity but in whether it could possibly be the mother species for all of the australopithecines. If *anamensis* was the common ancestor to both *africanus* and *afarensis*, then my theory that these latter two were sister species would be significantly stronger.

On the way from the airport into town, Meave recounted the story of how *anamensis* was found. She had had a hunch that

Kanopei, at Lake Turkana, was the place to look after a very old hominid jaw fragment had been found there in the 1960s. The museum's Hominid Gang was sent into the area, and there they found fragments of bluish fossilized teeth protruding from a rock. The teeth were hominid and were extremely primitive. A few days later they found a tibia nearby, and the level of excitement in the camp began rising.

My ears pricked up at the mention of the word tibia. This little knee bone is a key indicator of how comfortable its owner would have been walking on two legs. It was also a crucial component in assessing whether *anamensis* would have had long arms and short legs—what I would have hoped for in an ancestral australopithecine. Meave continued her story. After this discovery, she had reluctantly returned to Nairobi for administrative commitments, when she got a radio call from the excavation team leader, Kamoya Kemeu. They had found more teeth. She made the day-long drive back to Kanopoi as quickly as she could, negotiating the badly rutted roads that led through the mountains to northwestern Kenya. On her return, they marked out an excavation area and began systematically sieving the loose soil and smaller rocks. Gradually, more and more teeth emerged until they had an almost complete lower jaw set in almost perfect condition.

On the last weekend of the digging season, Richard Leakey flew up to spend the weekend with her. He'd lost his legs a year earlier in the aircraft crash, but his enthusiasm for flying—and for fossil hunting—appeared undiminished. His struggle to master the art of walking with prosthetics had earned him a great deal of sympathy and respect from the Hominid Gang. Meave described how one of them, Peter Nzube, had shouted excitedly in Swahili from the excavation site he'd been working: "Come quickly, it's wonderful." Accompanied by everyone in hearing distance, they followed him down the path that they'd all often walked on before, and there, sticking out of the sediments, was a complete lower jaw and next to it part of a skull.

As we drove through the streets of Nairobi, Meave told me that from initial analyses it appeared that *anamensis* was extremely primitive, at least as far as the skull was concerned. It had the largest canines of any hominid species yet described and had almost ape-like premolars. Combined with other features of the head and jaw, *anamensis* looked like a good common ancestor for all later hominids. But true to the pattern that I'd been experiencing, the postcranial bones told a very different story.

It is a remarkable thing about these early hominids, how the skulls and the rest of the skeletons almost always seem to be at odds with each other. The particular problem in this instance was a distal humerus that had been found years earlier, now ascribed to *anamensis*. It was large and somewhat confusing in its morphology, with those who had studied it at odds over how humanlike it was. If *anamensis* had a primitive head and human-like upper limbs, then it would have been a good candidate for giving rise to *afarensis* but not to *africanus*—unless the lower limbs were unusually small, in which case it would have had long arms and short legs and therefore would have been the perfect ancestor. Terms such as "humanlike" were becoming meaningless to me unless there was a sample of both upper and lower limbs. Nonetheless, I had the casts of my best fossil friend, Stw 431, with me, and at the very least it would be fascinating to compare Meave's bones with mine.

Of greater and more immediate interest to me, however, was that tibia. I had become the temporary tibia king of Wits during an intense period of research in which I'd studied every example of this little knee bone, from the most primitive hominids available to those of modern humans. A very primitive and complete *africanus* tibia had been discovered at Sterkfontein, and Tobias and I had recently published our findings in the *American Journal of Physical Anthropology* under the title "A Chimpanzee-like Tibia from Sterkfontein," highlighting in the paper the very ape-like morphology of the little hominid tibia. Little did I know at that time the fuss that would accompany that seemingly innocuous description of the small knee bone. The tibia is obviously a critical bone in helping anyone walk upright. In a biped during each step, almost the entire body weight is transmitted through the knee joint. Thus, if you are looking for clues to bipedalism, the structure of the knee is a good place to start.

Bipedalism, however, was of secondary importance in my quest to ascertain the size of that tibia and how primitive it was in relation to *africanus*. I had seen illustrations of Meave's tibia and was surprised by how large they appeared. Long arms and long legs were exactly the opposite of what I'd expected in a four-million-year-old hominid, which made me doubt whether *anamensis* was the mother of both *africanus* and *afarensis*—that was if Meave's tibia *was* in fact from *anamensis*, which in turn *was* 4.1 million years old. It stood in such marked contrast to all of our *africanus* tibia fragments from Sterkfontein, which were tiny. Even though we only had five specimens, the biggest was maybe only half the size of the *anamensis* tibia, if the drawing I had seen reflected the size accurately. It seemed to be even bigger than the *afarensis* tibia from the First Family site in Ethiopia. What also made me deeply uncomfortable, though, was how derived or modern in its morphology the *anamensis* tibia seemed in the illustrations. If that was the case, then, *anamensis* could not have given rise to *afarensis* either. The dynamics of early hominid postcranial anatomy were not only challenging the orthodox human family tree, but they were beginning to confuse the hell out of me as well. How could a creature that was supposed to me more ape-like in its cranial morphology than *africanus* and *afarensis*, and had lived at least half a million years earlier, be more advanced in terms of anatomical evolution from the neck down?

We arrived at the Kenya National Museum, its bronze of Louis Leakey sternly guarding the entrance to the "working" aspect of the museum, and walked through the building straight to the fossil-hominid strong room. There we were met by Emma Mbua, the curator of the collections. She led us into the inner sanctum. With a glint in her eyes, Meave pulled out three drawers of fossils. Picking up a new *anamensis* mandible, she handed it to me. It was clearly one of the new finds.

"What do you think?" she asked.

I looked at the specimen jawbone and hefted it gently in my palm.

"The canines are pretty big," I said, mentally comparing it with the sample of other early hominids, "but the premolars are the most convincing of all. They sure are primitive."

"I know," Meave said leadingly, "and what do you think about that accessory cusp on the canine?"

Such a cusp is also usually considered primitive among hominids. I examined it. The morphology looked suspiciously like the smaller *africanus* specimens we had in our safe back in Johannesburg. And then I realized that of course this feature wasn't present, or at least not common, in *afarensis*.

"Well, its certainly different from anything I've seen in the Hadar collection," I offered, "but I wish that I had brought some of the dental specimens of *africanus*. It looks a bit like some of ours."

Meave looked at me quizzically before wondering aloud: "That would be strange for it to be in *anamensis* and *africanus* and not in *afarensis*."

"Wouldn't it just," I confirmed. This was getting complicated. On the one hand, the fossils we'd been examining suggested that the *anamensis* skull could have given rise to *africanus*, although the initial descriptions of the postcranial elements indicated that *anamensis* gave rise to *afarensis*. My levels of intrigue were rising.

"Now, where is that tibia?" I asked, getting curious and impatient. Meave reached into another drawer and pulled the specimen out. I was immediately struck by its size. "Its bigger than I thought," I commented as I handled the two pieces of fossils, turning them over in my hands. "If I didn't already know where this was from, I would have guessed Koobi Fora at about 1.7 million years and not 4.1 million years."

"Why is that?" she asked.

"First, it's the size," I said, holding the proximal and distal parts of the specimen together. "This fellow is not only big, its long." I put the two halves in approximately the position I thought would represent their total length. "Second, the morphology of this proximal part of the tibia is derived. Look how the condyles are not only flattened, they're depressed." I pointed to the two proximal joint surfaces, running my fingers over their surfaces to illustrate my point. "Also, the attachment for semi-membranosus, a ligament attachment point, is elongated not rounded as I would expect in a tibia from this time period. Third, the shaft is long and straight."

Anamensis was displaying a tibia that seemed to be far more advanced than the hominid species that came after it. It did not make sense that a hominid so far back down the time line of bipedal history should have been able to walk better than descendants could a million years later. To illustrate my point, I turned and reached into my daypack.

"Now, let me show you something." I pulled out the cast of the little Stw 517 tibia from Sterkfontein, Member 4. This specimen belonged to an *africanus*, a hominid that lived only about 2.5 million years ago—more than 1.5 million years younger than Meave's tibia. I handed her the cast.

"See what I mean. This one's as small as Lucy. Now I know it might be a female, but that would mean that all of the tibial specimens we have from Member 4 are female. That might also be, but I'm seeing the same 'small' pattern in all the postcranial elements and I doubt everything is female." I took the head of the Stw 517 tibia back from Meave and held it against her *anamensis* tibial head. "See what I mean about the morphology? This one from South Africa has a curved lateral condyle and rounded semi-membranosus attachment. All these are primitive characters, and yet Lucy doesn't have all of that even though she's small." I put the *anamensis* tibia down and ran my fingers down the shaft of the Sterkfontein tibia. "Furthermore, from this you can see how curved the shaft must have been. Even only with a few centimeters of bone, you can see its turning away. This thing was going to be curved like an ape's." Meave took the cast back and examined it carefully.

"It certainly does look different."

"Could your tibia be from something younger?" I asked, looking again at the specimen from Lake Turkana. "Maybe its from a younger horizon that has eroded down to older beds, or even a burial?" Meave shook her head, puzzling over the differences between the specimens.

"I don't know, but I'll check."

I didn't see Meave the rest of the afternoon. Emma Mbua set me up in a visitor's room and I got to work on the shoulder-girdle elements in the collection, checking the published measurements against my own and examining the fossils for morphology that wasn't clear on casts I had back at the labs in Johannesburg. I also had a long look at the magnificent Turkana Boy skeleton and KNM-ER 3735, the partial skeleton that was supposed to show some indications of having long arms and short legs. I was dreadfully disappointed. The skeleton was little more than scraps of bone. If one wanted to wish morphology into the specimen, you might argue that it had very robust shafts in the upper arms and was more gracile in the lower limbs, but even that would be stretching the limits of imagination.

That evening at my hotel, sitting on the veranda and having my first Tuskers beer in almost five years, I thought about the material I had seen that day. It was becoming very clear that there were two completely different regional evolutionary patterns emerging. The East African one was distinguished by the early hominids having ape-like skulls and very humanlike limb proportions at an early stage in their evolution. Then in South Africa, a phenomenon that was virtually the opposite with *africanus* displaying a more humanlike, or advanced, head and yet retaining very primitive body proportions until quite a late stage in the overall hominid evolutionary picture.

I sipped the malty brew and wondered how to connect the two processes. To do so would mean accepting that evolution went backward for a period, but I struggled to find the logic in early hominids starting to develop back into ape-like bodies while their heads were evolving toward the early *Homo* form. If I was to rely on logic—and there's not necessarily any comfort in logic from such a distant perspective of our past—I would have to accept that hominids in the two regions evolved entirely separately from each other. In turn that would mean that humanlike bipedalism, in the form of humanlike body proportions, had arisen twice in Africa— that there were two entirely different models for how we first started walking on two legs. If so, where does the ancient lineage of *Homo* lie? East or South? I didn't see how both processes could have led to the same genus. I decided that I would show Meave the postcranial problems that were emerging in Stw 431 and get her opinion. Maybe she could shed some light on the complexity.

The next morning I worked for a couple of hours in the small museum office before Meave made her entrance. She had also been thinking about the implications of the issues that had arisen the previous day. She told me she had checked the dates of the fauna from around her *anamensis* tibia and that the integrity of its age was undoubtable.

"There's nothing there at all to indicate contamination from a younger deposit," she said.

I was impressed that she had taken my case seriously enough to double-check the entire faunal sample. So there was no disputing that the modern-looking *anamensis* tibia was actually much older than the *africanus* one.

"I don't know, then, Meave. The whole thing puzzles me. I've also got some other weird things popping up in the postcranial sample that I'd like to go over with you."

Meave smiled. "I've got a better idea," she said. "Richard's busy today, but come by at lunchtime tomorrow. He'd love to see you, so why don't you give us all here a slide show and show us what you're up to down in South Africa."

The next day I gave the Leakeys the long arms–short legs slide show. I outlined the work that McHenry and I had done, and then showed them the casts of Stw 431, in particular showing them how the ulna and humerus looked so large against the pelvis and the sacrum. Richard Leakey's first response was studied disbelief.

"Okay, I see that it's got unusually big arms for that pelvis and sacrum, but how do you know it's from one individual? You know those South African caves. Sterkfontein's just a big jumble."

He had a somewhat smug look on his face because he knew he was making a good point. Recently he'd visited South Africa, and I had personally shown him the location at Sterkfontein of the Stw 431 find. On that occasion I had raised the problems of South African cave stratigraphy, and how extremely difficult it was to associate even those remains that were found side by side.

"Well, two things argue strongly in favor of this skeleton being one individual," I responded, picking up the ulna and using it as a pointer to illustrate my argument. "Alun Hughes told me that the body came out in an anatomical position, shoulder and upper limbs at the top, vertebral column in the middle, and pelvis at the other end. Plus, all of these vertebrae articulate perfectly. Secondly, the rest of the postcranial elements from the site show the same pattern of large upper limbs and small lower limbs. We don't have any lower-limb elements anywhere as big as the big Hadar individuals; all of ours are Lucy-size or just a bit bigger. The upper limb sample from Sterkfontein is just the opposite: All the humeri, ulnae, and radii are big, as big as the biggest *afarensis* ones, and none are small."

"How many individual adult postcranial elements do you have?" Meave interjected.

"Almost 50 adult upper-limb and lower-limb elements," I answered, thinking I was winning a major point in the debate. "Besides," I went on, "it shouldn't be too surprising to see an *africanus* with long arms and short legs." I turned to Richard. "You and Alan Walker pointed out that KNM-ER 3735 and OH 62 had long arms and short legs in *Origins Reconsidered*.

"Yes," Richard said, nodding his head yet indicating that he remained unconvinced. "But those South African caves...."

My discussion with the Leakeys made me realize that the biggest handicap I faced in getting acceptance for the long arms–short legs theory was not the actual physical evidence. It was the fact that the South African sites were so complicated and so difficult to date that made most scientists reluctant about basing any kind of evolutionary theory on such shifting sands. Paradoxically, this was quite a comforting insight as it tempered my impatience to have the long arms–short legs theory immediately recognized and accepted. It also made me more determined than ever that I had to persevere with showing that *africanus* had such different body proportions from *afarensis* as to force a revision of our understanding of human evolution. But I had to be aware that the tumble-drier nature of South African fossil formation was going to be a useful weapon in the arsenal of anyone wanting to cast doubt on the validity of this theory.

The notion that the bodies of *africanus* and *afarensis* were similar in morphology was a basic assumption in the post-Lucy orthodoxy, which held that bipedalism had arisen very early in human evolution and that the heads took a long time to evolve. The implications of the new interpretation of *africanus* were that the heads had developed earlier on and that humanlike proportions had arisen either twice (in an *afarensis* to *africanus* phylogenetic scenario) or relatively late (in an *afarensis* and *africanus* as sister species scenario) in human evolution. But both views could not be accommodated together in the existing orthodoxy. Specifically, if *africanus*, with its more derived cranial and dental features and its ape-like body proportions, could be shown to be ancestral to species like *Homo habilis*, or just more similar, then the position of *afarensis* in the human family genealogy was under siege.

On the flight back to South Africa, I realized that if I was going to stick by the idea that *afarensis* and *africanus* were sister species with a missing common ancestor, it was going to be viewed by many as if I was throwing Lucy out of the family tree. I would be naïve to expect anything other than vociferous opposition from the most powerful people in paleontology. There were many leading scientists who had invested a great deal of time and energy in promoting Lucy and her ilk as the trunk of the human family tree, and they would not take kindly to the morphological challenge that *africanus* was mounting against their views. I settled back in my seat and looked at the clouds. The weak link in my argument was the difficulty in securing reliable dates. I had to find some way of proving that if *africanus* lived at the same time as *afarensis*, that alone would bolster my sister-species argument, which would in turn help me convince McHenry that his evolutionary reversal theory was wrong. I would also then be ready to take on the Great White Shark himself, Tim White, nicknamed as such because of his aggressive intellect and inquisitor's mind. White is probably the most formidable personal force in paleontology today, having been involved in the description and analysis of the world's most important fossils during the past 25 years. He is also credited with finding the oldest known hominid, *Ardipithicus ramidus*, dated to 4.4 million years ago. I knew that the point at which the long arms–short legs theory was being taken seriously by the establishment was the day that Tim White set his sights on me.

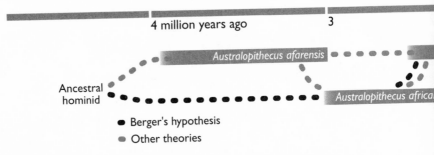

The "linear model" of the evolution of the hominids based on the long arms–short legs theory illustrates the shift in body proportions from short legs (primitive) to long legs (derived) to short legs (primitive) to long legs once again. This model is supported by Henry McHenry; the author believes that it is more likely that there was a division early in hominid evolution that led to two separate experiments.

Dating is one of our biggest problems in South Africa. Pinning specific dates on any of the South African sites is a hazardous process, and so we are forced to rely excessively on faunal comparisons. By comparing the animal remains found in association with South African hominid fossils with those from the more reliably dated East African sites, we can "bracket" the South African sites within broad time periods. This still poses problems of complete accuracy because there are no identical faunal collections in the two regions. Therefore dating is continually an issue in the discussion around South African fossils. Take Makapansgat, for instance, which, along with Sterkfontein, represents the oldest hominid site in South Africa. Tim White argues that the dates of Makapansgat are probably just shy of three million years old; whereas Tim Partridge of Wits, using paleomagnetic dating to examine evidence for periodic reversals of the Earth's magnetic poles, and Kay Reed of the Institute of Human Origins, using comparative faunal dates, would argue that they are closer to 3.4 million years. One has to accept that White may have a vested interest in his dating, but so might Tim and Kay. If Makapansgat is indeed 2.9 or 3 million years old, it would place it neatly at the termination of *afarensis* in East Africa. If the older dates are correct, however, that means that *africanus* and *afarensis* were contemporaries.

Present

Later *Homo* species including *H. sapiens*

Homo habilis

Homo erectus

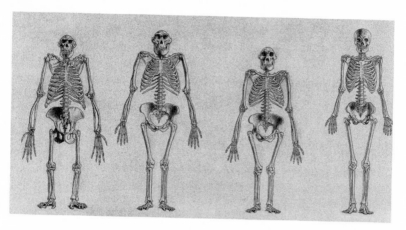

Illustrations by John Gurche of the skeletons of apes, hominids, and humans demonstrate the extremely long arms in A. africanus and the humanlike proportions of A. afarensis—from left to right, a male chimpanzee, a male afarensis, a male africanus and a male human. Note particularly the position of the wrist against the thighbone.

The signal that I might be becoming a threat to Lucy arrived in the form of a group of scientists led by the Great White Shark himself. It comprised Berhane Asfaw, a leading Ethiopian scientist; Owen Lovejoy, the king of postcranial anatomy, and probably the leading terrestrialist; F. Clark Howell, one of the doyens of archeology and paleoanthropology; Scott Simpson, a dental specialist; Bruce Latimer, the postcranial prince and Lovejoy's emerging heir apparent; and Gen Suwa, a leading Japanese dental anthropologist. They had come to South Africa to compare the newly discovered *Ardipithecus ramidus* material from Ethiopia with the South African fossils. These extremely unusual ape-like animals had been announced as the oldest hominid fossils yet discovered (4.4 million years), and had first been put into the genus *Australopithecus* by Tim and colleagues. Then, in a brief corrigendum, White and colleagues had created a new genus, *Ardipithecus*, citing the extremely primitive and unique nature of the morphology of the handful of specimens they had announced. Rumor had it that they had uncovered dozens more specimens of this ape, including post-cranial bones and a whole skeleton. Being so much earlier than anything yet described, and looking very apish from the little we in the scientific community could see, the whole paleontological world was waiting for the description of this material.

I would also later learn that White and his team had discovered another hominid in Ethiopia, one they hadn't yet told anyone about. It would turn out to be a very odd hominid at 2.5 million years that was revealing an unusual pattern in limb length, so the South African material was extremely pertinent to their study. Nevertheless, the graduate students in the department were so impressed with the status of these scientists that the group was quickly labeled the Dream Team after the U.S. Olympic Basketball team of years earlier.

The Dream Team arrived in style. They touched down at Johannesburg International Airport in a private 727 jet belonging to Anne and Gordon Getty, who are dedicated supporters of Tim White's research efforts. They swept in, literally, like rock stars. They brought with them all their own equipment, including chimp and gorilla skeletons as well as human casts. The team set themselves up in the Wits fossil lab, commandeering the long green felt table that occupied the center of the room. They turned it into a bizarre fossil assembly line, so that they could go through every piece of bone in our collection very rapidly in just a few days. Their visit prompted Meave Leakey to fly in from Nairobi to personally see the latest South African material herself and to get a chance to compare her *anamensis* material with *africanus* and *ramidus*. I was charged with looking after this group of luminaries as Professor Tobias was out of town.

White and I get along very well, although I sometimes get the feeling that he sees me as a naïve youth who will eventually find the true path—in other words, see human evolution the way *he* sees it. I've known him since the early nineties and have always been struck by the intellectual power that instills a fierce loyalty in his team, despite his reputation as a real taskmaster.

The Dream Team's arrival coincided with South Africa's emergence from the paleontological backwaters of the apartheid years. I have this image of South Africa as a finned lobefish struggling out of the water into the mud on its journey toward amphibianhood. The South African perspective on human origins was beginning to get international publicity. NATIONAL GEOGRAPHIC had just covered some of my research in which I was championing *africanus* as having played a more central role in human evolution. Clarke and Tobias had announced the discovery of "Little Foot," a series of connected foot bones from Member 2 at Sterkfontein, which was being billed as the transition form from ape to human. Tobias and I had authored the paper "A Chimpanzee-like Tibia from Sterkfontein," pointing out how much more primitive this tibia was than any comparable anatomical element in *afarensis*. Henry and I had finished a draft of the long arms–short legs paper and were just about ready to submit it. All this was beginning to resemble a concerted challenge on the predominance of what I'd call the East African hegemony of human evolution.

It was becoming clear to White and the Dream Team that the direction in which we were heading implied that *africanus* was more ape-like than *afarensis* from the neck down, and was thus challenging Lucy's mother-species status. This was not a theory they were going to take lightly. It was the beginning of a challenge, not only to their intellectual authority but to their reputations as the world's leading evolutionary scientists. White was the co-architect of the Lucy-as-mother-species theory, while Owen Lovejoy had personally developed the paradigm in which *afarensis* was seen as humanlike from the neck down.

Their strategy in dealing with the challenge became evident on the second to last day of their visit. They were going to tackle each of the *africanus* issues individually. Once they had finished looking at all the material and had conferred among themselves, they called me in to the fossil lab and asked me to sit down. I felt uneasy as I took up my chair. It was judgment day. I braced myself with the image that *africanus* was my client and I was his attorney hired to defend him at all costs. Bruce Latimer pulled up a chair on the other side of the table opposite me. He had the air of a determined prosecutor intent on putting someone away for a long, long time. It became clear that the Dream Team had rehearsed this scenario. Latimer was charged with convincing me that my interpretation of the fossil record was incorrect. F. Clark Howell sat at the end of the table to my right; Scott Simpson beside me; Owen Lovejoy sat next to Latimer. White, who had automatically assumed the air of chief inquisitor and hanging judge, spent most of his time pacing behind me. Berhane Asfaw and Meave Leakey sat at the other end of the table, a look of bemusement on their faces, as if they were in some ancient Roman gallery watching some poor wretch being fed to the lions. My unease turned to terror as my self-image of *africanus*'s defense attorney slipped. I began to feel more like a heretical young priest appearing before an inquisition trial in the Middle Ages.

The proceedings began. On the table before me they laid out a variety of comparative material. White called for the cast of the little *africanus* tibia (Stw 517) that had been the subject of my paper, and asked if he could draw on it. I was surprised but agreed. He took out a felt-tip pen and drew on the areas that were not preserved, that had been broken off in the fossilization process and on which he felt one couldn't make morphological assumptions. He sketched muscle attachments on the cast in black. They then began to work on me.

Their argument was this. I had done the science a great disservice by describing the tibia as chimplike. This implied that it was more primitive than *afarensis* and therefore could not have evolved from it. Taking turns to drive their point home, they maintained that the tibia fell within the range of variation found in *afarensis*, and this being the case, meant that it was closer to being humanlike rather than chimplike. At worst it could be called *afarensis*-like. "Chimpanzee-like" was beyond the bounds of acceptability. Having stated the case against me, they pulled out the *afarensis* material they had on hand and began pointing out the features. There they all sat, with their boxes of world-traveling, nonhuman, primate material at hand, all their casts and my little tibia, and asked me to go through my case point by point.

Taking a slow, deep breath, I began my argument. My defense was going to rely heavily on the technical interpretation. Tobias and I had labeled the tibia chimplike on the basis of four main points. First, the lateral condyle—the platform that the outside of the base of the femur rests on—was extremely curved, more so than you would see in a human, and therefore represented greater mobility in the joint. Second, it had a circular attachment for semi-membranosus on the side that tends to be associated with ape-like characteristics. This was based not only on our own observations but also on a body of convincing work that had been built up by a group of scholars including Brigitte Senut and Michelle Tardieu, French functional morphologists who had developed a theory that this was largely absent in humans but present in apes to assist in greater mobility and stress resistance. Third, from the small bit of shaft that was preserved, we deduced that the whole of the shaft was more curved, or ape-like, rather than straight as one would find in humans. Finally, there was the distal aspect: There was a small piece of the distal articular surface, which is the bottom of the tibia, in which we felt that the angle of this joint was slightly offset, sloped as in apes rather than being relatively flat as found in humans.

They agreed that the lateral condyle was curved, but that it was not *much* more curved than some of the *afarensis* tibia, particularly the smallest examples of this bone from Hadar. I took their point but replied that this tibia had a greater absolute curvature than any fossil they had found. They had four tibia and we had one, and our one *was* more curved. We debated this over and over. It is just a *little* more curved than Lucy, they maintained. "Yes," I concurred, "that's exactly the point. It *is* more curved than Lucy's. If this little tibia represented the norm in *africanus*, then all the tibia would be more curved." Owen Lovejoy then picked up on a new tack. He pointed out that his own work in studying Native Americans had proved that the tibia could be a highly variable part of the modern human anatomy. If this was the case now, it could have been the case two million years ago. I conceded there was an argument for variability but pointed out that this was the tibia of a single individual, and as such we had to presume that it represented the normal morphology of this species. One certainly wouldn't expect to find curved lateral condyles in the average human, I said, not even in his variable Native American population. A half-hour discussion on the issue of variability ensued, and when it appeared that neither side was going to back down, we agreed to hold off on a conclusion for the moment.

Then the Dream Team moved on to the next piece of evidence: the attachment of semi-membranosus. Bruce began the discussion by arguing that at least two of the Hadar specimens have a circular semi-membranosus attachment, that these are smaller specimens, and that the larger specimens do not. I acknowledged that there was a variation in the Hadar group. But, I argued, when one combined the curvature of the lateral condyle, which *was* greater in the *africanus* tibia than in anything at Hadar, with the rounded semi-membranosus attachment, one had to presume—given the small sample from *africanus*—that this was going to be the dominant characteristic for the species. What I didn't bring up, but what Tim White was well aware of, was that Senut had used this difference in morphology to argue that there was not one but two species of hominid in the Hadar fossils. This, I knew, was not the time or place to bring up that sticky issue.

White wasn't buying my line of reasoning. He stopped his restless pacing behind me and leaned over my shoulder to speak. "Look," he said, "we can show you each feature in a specimen from Hadar."

"I know," I replied, "but you can't show me a single tibia from Hadar with all these features in one specimen."

"Well, we'll find one," he said impatiently.

"I don't think so," I replied, feeling my confidence levels beginning to rise.

We then moved on to the curvature of the tibial shaft. This would be significant because humans tend to have very straight shafts, while in apes they are much more curved. I got the group to acknowledge there was curvature, and then appealed to their anatomical prowess, saying that they had to see that the curvature *was* very chimpanzee-like. To emphasize my point, I reached into the lab desk drawers and pulled out two female chimpanzee tibiae from our comparative sample. I held them next to the Stw 517 tibia and showed how they were in fact less curved than the *africanus* tibia's shaft.

"*That's* not a chimpanzee tibia!" Latimer was becoming a little more aggressive. He stood up and moved swiftly over to one of his boxes that the team had brought from America. He reached in and pulled out a bone. "*This* is a chimp tibia."

Clearly, it was from a large male chimp. It was curved along the shaft, much more than my two little female specimens.

He added dismissively, "Yours are silly little chimps."

"But they are chimps nonetheless," I insisted. Round and round we went with a fresh debate about variability within species. The whole group took turns in trying to convince me that the variation in the Stw 517 tibia was only *slightly* greater than that seen in the Hadar sample, and therefore could not be imbued with any significance. I began to get frustrated. They were so intent on driving home their argument that they weren't actually listening to me. The whole point was that no single individual at Hadar showed all of these features. That was my only point, and no matter how much one argued the toss, this fact was clear.

We finally moved on to the distal end of the specimen. Latimer pulled out an index card and laid out two human tibiae and several chimpanzee specimens, including my own, on the table. I sensed trouble coming.

"Let me show you something," he said, holding one of my chimp tibiae upside down in front of him. "If you want to look at curvature of the distal articular surface, all you have to do is this." He placed the edge of the card along the distal surface, clearly showing the high angle created by the card's length. He then picked up the human and performed the same measure. The card was nearly horizontal. Picking up Stw 517, he held the card to the end. The card lay nearly horizontal. In one elegantly simple experiment, Latimer had convinced me that we had, for some reason, over-judged the angle. But I wasn't going to let my argument collapse like a house of cards: "Ok, I'll give you that one, but that still doesn't convince me about the proximal end. It just means they are bipedal, that's all.

Everyone was beginning to tire from the intensity of the concentration. We'd been talking now for over an hour. It was time to sum up. I was feeling more like *africanus*'s defense lawyer again rather than a sacrificial victim. I had also figured that I could concede the point about semantics without my case as a whole being compromised. I imagined myself standing before the proverbial judge: "Your Honor, I plead guilty to the lesser charge of being too loose and liberal in my application of the word 'chimpanzee,' But, sir, I feel that this does not affect the standing of my client for the following reasons...."

Instead, I restated my case as follows: Although you could find any individual character of the Stw 517 tibia within the Hadar sample, no single tibia in the Hadar sample carried the whole suite of primitive characteristics that Stw 517 had. Therefore, ours was a more primitive morphology than that exhibited by the east African specimens. That was the best hypothesis we could make given the single specimen that we had.

I was surprised to note how much tension there was in the room. I paused before continuing, saying that I did acknowledge that "chimpanzee-like" may have been the wrong title for the paper, given the equivocal nature of the distal end. It wasn't truly chimp-like even though it was similar to the small chimps we had used in our study, but not as pronounced as some of the samples they had. Perhaps we should have said, "a primitive early hominid morphology in a tibia" or emphasized how this tibia was unique and unlike that of any living large-bodied primate or fossil yet found.

That was the sign that White had been waiting for. He immediately ceased his pacing and came over to me. Pulling his pen out of his pocket with a slightly over-exaggerated flourish, he put it on the table in front of me with a copy of my "chimpanzee-like tibia" paper.

"Sign it and say you retract the title," he ordered in a camp commander's voice, although I did notice the hint of a smile playing around his lips. The mood of the whole group took its cue from White and lightened, despite the seriousness of the debate. After all, the question we were grappling with was of consequence to every living being on the planet. A few smiles emerged from around the room as Tim drew himself up to his full height and indicated with a tilt of his head that he wanted me to sign. In a cracked voice, I put on a bad Spanish accent, parodying the inquisition I felt I'd been through, and said, "I'm sorry, sir. I cannot sign this confession. All the fingers in my hand, they are broken." I never did sign the paper.

CHAPTER 9

HOMINIDS ALIVE

I felt a sense of triumph after taking on the Dream Team. By the latter part of that year, the body-mass paper comparing the relative size of the upper and lower limbs went into the *Journal of Human Evolution*; we held the limb-length paper for the *South African Journal of Science*. Henry and I sat back to await the referees' comments. I was sure that I was on a winning streak.

In June 1997, Henry and I received the reviews from the editor of the *Journal of Human Evolution* for our long arms–short legs theory. They were surprisingly positive about the article; although they balked at some of our conclusions, they did not challenge the thrust of our work. We made the minor corrections that the referees had suggested, argued to keep some of our other points intact, and submitted the final manuscript for publication.

The body-mass paper was accepted for publication in September, and we were informed that it would take about a year to appear in print. We had already prepared the limb-length paper and submitted it to the *South African Journal of Science*, a journal with a faster turnaround time. Both of the manuscripts would appear in print simultaneously.

By then I had shown the research to Bill Allen, editor of NATIONAL GEOGRAPHIC, and he expressed interest in having me write an article about the research, which would be published around the same time as the two papers. I spoke with the editors of both journals and they agreed to hold the papers until August 1998, the earliest that the GEOGRAPHIC could print the article. After confirming that the journals were willing to work with us, I requested that John Gurche be appointed the artist for the project. He and I had worked well together on the earlier NATIONAL GEOGRAPHIC "Dawn of Humans" series, and I felt that he would project my concepts better than any other artist.

Over the past year, I had also been involved in organizing a large international congress to be held in South Africa in 1998. It was to be Tobias's swan song. Literally hundreds of top academics from around the world were coming to South Africa to join in the largest conference of its kind ever held in Africa. The Dual Congress arrived in July, and as expected almost 700 delegates from 70 countries came. It was truly exciting, even though its location at the Sun City entertainment complex was somewhat bizarre. Sun City is two and a half hours out of Johannesburg toward the Botswana border. It was conceived of as a revenue generator for the apartheid homeland of Bophuthatswana, and it is a startling sight. After driving through a number of littered, poverty-stricken rural communities, one rounds a corner of the Pilansberg mountain range to find this tropical mirage rising out of the bushveld like some sort of Vegas casino. Nonetheless, Sun City has survived the legalization of gambling in the rest of South Africa by turning itself into a family-oriented entertainment center and conference venue. Almost every active scientist in paleoanthropology and human biology attended. The fake African kitsch décor and the bushveld Disney-like ambience of the center provided a surreal setting for the debates and discussions that filled six days of intense scientific interaction.

Unfortunately, Henry McHenry couldn't come, so it was left up to me to read a paper on our limb-length results. Since this and the body-mass publication were due out imminently, we felt that the conference would be an opportune time to air our views. I was also asked to deliver a paper in Meave Leakey's symposium on early hominids, and I chose to review the extensive *africanus* postcranial material we had. Also in the colloquium was Tim White, who, in addition to discussing the new *ramidus* material, dealt with specific and general trends in Pliocene hominid evolution.

White's paper was given before mine and he made a superb presentation, taking a jab at what he called "X Files" paleontology, a term borrowed from the popular fictional investigative TV series dealing with conspiracies and the supernatural. In his laconic and intellectually rugged manner, White accused certain researchers of formulating hypotheses without data, in particular the "hard" fossil evidence. What he was saying, in effect, was that he who has fossils gets to do the science. I felt this was a little bit harsh, since White wasn't allowing anyone but his team to work on his *ramidus* material, and one had to expect that people would speculate on the little that had been published. Nevertheless, near the end I was surprised to see him conclude with a discussion of evolutionary trends in the Pliocene and early Pleistocene. He stated unequivocally that the trend was obvious, while there was clearly a speciation event after 2.5 million years, the Pliocene was characterized by the existence of *only one* hominid species *at any one* time. First there was *ramidus*, then *anamensis*, then *afarensis*, and then *Homo*, he concluded, in what seemed to me to be a dogged clutch at the old ladder image of human evolution. What about *africanus*? I thought to myself. Does it not exist? I realized that White was not at all going to like the results that Henry and I had come up with on long arms–short legs.

After the applause died down, I stood up to do my presentation. I reviewed the substantial numbers of new fossils that we had for *africanus,* briefly mentioning the odd distribution patterns we were seeing, and said I would save the detailed discussion of that for the long arms–short legs paper I would be delivering the next day. At the end, I couldn't help myself and took a dig at Tim's argument that there had only been one hominid species at any one time in the Pliocene. I pointed out that this seemed to ignore not only *africanus* but the *barhagazeli* fossils from Chad. I noted in conclusion that I would probably be accused of venturing into X-Files paleontology, too.

The next day, during my presentation, White sat in the second row from the front. I knew I was in for a bit of payback. I was finished with my presentation and took questions. After fielding a few queries and comments, I noticed White's hand shoot up. He had that grin on his face.

"So, Lee, was that a *virtual* femur you showed us on the 431 skeleton?" He was referring to the fact that Stw 431 had no femur. Henry and I had extrapolated femoral-head size from acetabular size, a proven method with an accuracy rate of over 98 percent. We had then calculated femur length from this estimated head size. Femoral-head size has over a 95 percent predictability of length, so while the end result wasn't perfect, we felt it was still a pretty good estimate over the whole length of the femur. I explained this to Tim.

"So you created a virtual femur length from a virtual head?" he asked. People around the room chuckled. I knew when to quit, and I guess I deserved a bit of this since I had challenged him in the earlier symposium. For me, this had become more of a duel congress than the Dual Congress it was meant to be.

"If you say so, Tim."

Despite White's remarks, which were really about the potential of our imagination to distort paleoanthropological findings, one of the most interesting and enjoyable aspects of this science is trying to recreate the appearance of early hominids by blending our ideas with the facts. Almost all of the popular dinosaur movies filmed in the past few years have employed scientists. They work closely with artists in rendering as accurately as possible the now-extinct animal's appearance and actions. Inadvertently, Hollywood is helping science by nurturing an emerging class of paleo-artists who can make a full-time living out of visually re-creating the past.

One of the best of this new breed of science artists is John Gurche. We have worked closely before, producing early hominid images for NATIONAL GEOGRAPHIC articles. Gurche is probably best known for his artistic work on dinosaurs, having produced a set of U.S. stamps and numerous posters on this theme, as well as some of the dramatic images in Steven Spielberg's *Jurassic Park*. But Gurche's first love is for hominids. His professional approach is evident in the time he spends researching his subjects and in the thoroughness with which he goes through the academic literature, as dry and dull as it sometimes can be. He also insists on seeing and measuring the original fossils before putting pen to paper. Some of my most pleasurable moments in the lab have been the anatomical debates I've had with Gurche over how the Wits fossils should be visually interpreted.

This type of work may seem frivolous to some of my more staid colleagues, but its value should never be underestimated. The impact that a single well-drawn or -painted illustration can have on the public—and even the scientists—is immeasurable. It brings to life something of the past in a way that words cannot, and certainly enables much of the rather specialized literature to be captured in a synthesized, accessible form. However, like the science, the images produced by these collaborative exercises are constantly evolving. Just how rapidly our ideas about what our earliest ancestors looked like can be seen in two of Gurche's illustrations, one created in 1991 and the other in 1998.

Reconstruction of a male A. africanus *walking, drawn in 1991, based upon the "old" model of an* A. africanus *body notes the similarity with the body of an* A. afarensis *as opposed to the "new" image of an* A. africanus.

Gurche came to Johannesburg in 1991 to see for himself the large collection of new material, in particular the large number of postcranial bones we had in the Wits safe. I was still a graduate student then, just a year and a half into my Ph.D., and immersed in the world of the functional anatomy of the shoulder girdle. John and I hit it off almost immediately. South Africa was still a state in isolation with very few international scientists passing through, and I was thrilled to have someone of his caliber to talk to about my research. He was also interested in postcranial anatomy, which is a rarity even among my colleagues. Obviously, this was before I had synthesized my thinking about *africanus* around long arms and short legs, so most of our discussions revolved around how muscular the upper bodies of these hominids would have been. All my research at the time showed how powerful *africanus* would have been when it came to upper-body strength. I could tell how strong the shoulder muscles were by the markings these muscles left on the bone—large raised areas, indicating how well developed they were. Gurche agreed with my assessment that *africanus* was still very dependent on an arboreal lifestyle, having developed similar ideas while working with the *afarensis* fossils in East Africa.

Our model hominid for the upper limbs and pelvis was the new Stw 431 skeleton because it was one of the best preserved specimens in the assemblage, and one with which I was very familiar. For a head we used the newly discovered Stw 505, the largest *africanus* skull that had been found. Our choice was determined by the massive size of Stw 431's shoulders—any of the other skulls, according to Gurche, would look microencephelatic, or small-brained, atop that massive torso. The end result of this first collaboration is an image of a hominid that is enormously well muscled in the upper and lower body and has limb proportions closely approximating that of a human. This was based on the assumption at the time that *africanus*'s body would have reflected Lucy's proportions. Also, we never really studied the legs, since South African lower-limb material at that stage was very fragmentary and practically unstudied. Thus Gurche's 1991 large male *africanus* supported the idea that the early hominids had humanlike bodies. There were of course differences, particularly in the hands, but at the time only the skulls seemed to differentiate *africanus* from *afarensis*. I was so pleased with this image that I chose it as the emblem for the embryonic Paleo-Anthropological Scientific Trust.

Less than a decade later, it is clear that the image Gurche and I worked so hard on, and put so much thought into, is incorrect. I hope, each time we do another series, that we get a little closer to the "truth" of what these ancient creatures looked like. In 1997 we had the opportunity to work together again on a reconstruction of *africanus* for an article I was writing for NATIONAL GEOGRAPHIC. Obviously, much had happened in the intervening years to change the way I viewed *africanus*. I had finished my Ph.D. in 1994, which gave me more clarity on the upper-body muscles; McHenry and I had completed our limb-proportions research and I had refined my long arms–short legs theory. Gurche had read proofs of the papers on our findings and had come to South Africa specifically to debate these points. Interestingly, we stuck to the original material that we used in the 1991 collaboration—the Stw

505 skull and Stw 431's limbs—but we factored in the new data. Gurche still insisted during his visit that we study every other postcranial bone from the Wits hominid vault as well. For a week we went through the numbers McHenry and I had come up with, relating these figures to the bones laid out on the lab table. Gurche constantly made sketches, challenged our calculations, and debated specific anatomical points until he was satisfied with the validity of our thinking. And even then he didn't stop. We spent the evenings at my house, sitting in the yard, gazing at the stars, and continuing the discussion.

A male A. afarensis *(left) walks beside a male* A. africanus *(right) in a 1997 physical reconstruction based upon the long arms–short legs theory.*

Eventually he had enough data from the preliminary sketches to produce his more detailed drawings. Over the next couple of months, I would receive draft sketches in the mail with an invitation to comment on and correct any aspects that I felt were inconsistent. The final product of this endeavor can be seen in the new image that Gurche produced for the article. What is striking is the position of the hand against the thigh, which illustrates not only that the arms of *africanus* are now longer, but that the legs are also shorter than we had originally believed. Indeed, the estimated position for the wrist of *africanus* against its thigh is very close to what one would predict for a chimpanzee.

So aside from offering us a better picture of *africanus* as a species, what are, in the end, the implications of this research? The jury is still out and there is a lot of work to be done, but some facts have emerged. First, we are going to have to completely reassess the accepted relationships of early hominids. The differences in body shape and proportions challenge the conventional view that *afarensis* was indisputably the mother species of all subsequent hominids. I remain convinced that it is extremely unlikely that a pre-*afarensis* creature like *anamensis* with a primitive hominid cranial morphology, but seemingly derived body, gave rise to a form like *afarensis* with a largely humanlike body that then evolved into the more primitive body of *africanus* again. This reversal does not make sense given the present fossil record. Something in this chain has to give, and it strikes me as the simplest hypothesis that *africanus* and *afarensis* must be sister species evolving from an as yet to be identified common ancestor. If it is eventually found that *anamensis* has relatively long arms, then it may prove after all to be this hypothetical mother species.

We do have at present a remarkable fossil record of early hominids, and it's well worth examining what that record is telling us about the look of the human family tree. The postcranial anatomy of the early hominids tells a different story about evolution than skulls. Much has been made about the development of teeth, that the reduction in canine size compared to that of today's apes is a sign of evolution. Let me ask this. What if the common ancestor of apes and hominids did in fact have small canines, and the larger canines we see in chimps are actually their own evolutionary feature? It is not logical to think that the hominid lineage had a sole franchise on evolution and that today's African apes have not undergone any physical transformation over the past five million years. We must remember that it wasn't a case of apes standing still in the evolutionary stakes and all the changes occurring only in human ancestors. Both creatures were faced with an environmental challenge and adapted to it in ways that suited their own particular habitats. Chimps are good at being chimps; hominids are good at being hominids.

Chimps and gorillas are adaptations of a large-bodied ape to tropical evergreen forests, while hominids are adaptations to everything else except evergreen forests. That still holds true today. You see humans scattered across Africa in nearly every habitable area except evergreen tropical forests, where it takes hyper-specialization to survive in that particular environment. If you're a human, either you have to have the pygmy-type adaptations—low population numbers and small body sizes—or you have to clear-cut the evergreen forests in order to survive. That's the fundamental difference between chimps—and their evolution over the past five to seven million years—and hominids and theirs. If this is the case, it may be entirely feasible that some features that we view as derived, such as small canine size, are in fact the primitive condition. In this case, we hominids maintained the inherited pattern of small canines, and it is an evolutionary development of apes that their teeth have become bigger.

Because of the large gaps in the fossil record, it is by no means clear how much of the universe of early hominids we actually know; there have been such a variety of new hominid finds that the relationship between them is confusing. As a result, and given that our overall sample is so small, any new hominid discovery has the potential to stand the entire science on its head.

So is the oldest hominid so far discovered, *Ardipithecus ramidus*, actually the right candidate for the mantle of first ancestral hominid? *Ramidus* might well be, given the dating estimate of 4.4 million years, but of course we'll have to await the description of this material. The problem is that the "first" hominid is going to be such a mishmash of primitive characters and generalized morphology that we may never be able to absolutely identify it. It's going to be an ape that uses bipedalism at least some of the time, but not necessarily all of the time. It will have some of the features of later hominids, but not necessarily all of them. The one problem with *ramidus* is that we have nothing to compare it with. At present it stands alone in time, a creature in isolation with no field of reference that we can use to make any form of comparison. This was precisely the same problem that Raymond Dart faced with the Taung Child, the first australopithecine. He simply had nothing to compare it to.

But given all this, *ramidus* appears to have been able to walk on two legs, although as Tim White wryly observes, it may have walked in a way unrecognizable to us as modern humans, stumbling around like some creature from a *Star Wars* bar. So let's start there: At 4.4 million years ago, there appears to be at least an elementary form of bipedalism. By the time we reach 4.1 million years, bipedalism and the basic fundamentals of hominid morphology are all there. This is clear from Meave Leakey's *Australopithecus anamensis* fossils. Certainly, the heads look like a good ancestral candidate for all later hominids. But for *anamensis* to fit my ideal of a "perfect" ancestral species of later hominids, it also must have the long-armed, short-legged proportions of *africanus*. If only it was that easy.

What is so frustrating about *anamensis* is the lack of postcranial information about it. Only a few scraps have been recovered, including the one single tibia that looks a bit large to me to be proportioned like *africanus*. But maybe it is. Certainly the few upper-limb specimens I've seen attributed to this species, namely a radius and a humerus, are very big. If I was a betting man, I'd wager that we will eventually find that at least the *anamensis* males are very large and do have long arms, making it a good common ancestor of both *afarensis* and *africanus*. But we have to wait for more discoveries to find that out.

In any case, if not this particular species then something very much like it spread throughout southern and East Africa, moving into almost any region that provided broken woodlands. As this species expanded its range, populations would encounter different stresses in different environments. Over time this would have led to a speciation event, in which different localized populations would have begun taking on the different morphologies best suited to their particular environments.

My guess is that at least two distinct creatures emerged from this process—*afarensis* and *africanus*. Why should they have been different? Well, if *afarensis* had concentrated on a more terrestrial lifestyle—that is, relying less and less on trees for shelter and feeding—it would have improved on its early bipedalism and become more adept at walking on two legs. In the course of this, a high degree of sexual dimorphism remained and the primitive head that it inherited from the ancestral form did not change much.

The other species, *africanus*, living largely in southern Africa, possibly remained close to a more forest-dependent lifestyle, climbing trees to sleep at night and venturing far less into the open territory than *afarensis* was comfortable with. This meant that it retained the more primitive body proportions, but changes took place to its skull and jaws in which both became bigger for reasons we still have to properly understand.

In short, Lucy may have been more adept as a terrestrial biped than *africanus*, whose shorter legs and more massively built upper body would have shifted the center of gravity and thus made their gait extremely nonhumanlike.

In essence, then, it's not so hard to view *afarensis* and *africanus* as fundamentally different adaptations of the same basic plan. In my opinion, they are as different as chimpanzees are from gorillas. Both chimps and gorillas knuckle-walk when they're on the ground, so they share a basic mode of locomotion. But they have fundamentally different bodies and completely different lifestyles. Gorillas are highly sexually dimorphic, live in small alpha-male dominated social groups, and by and large are vegetarians. Chimpanzees on the other hand are much less sexually dimorphic than gorillas, live in larger social groupings that have much greater social complexity, and enjoy a distribution in a much wider array of habitats. I would suggest that the differences between *africanus* and *afarensis* are as extreme as that between these two apes, with *afarensis* being in many ways closer to the gorilla model.

Where, then, does the complexity evident in the earlier australopithecines leave us in trying to trace the emergence of our own genus *Homo*? One of these species did *not* directly give rise to the genus *Homo*. Maybe neither of them did. But *both* could not have. To me, the cranial evidence and postcranial evidence available at this time suggest that *afarensis* almost certainly gives rise to the East African robust australopithecines. I draw this linkage through *Australopithecus aethiopicus* (the Black Skull), which morphologically appears to be ancestral to *boisei* (OH 5). *Africanus*, on the other hand, appears to be the most logical ancestor of the hominid we presently call *Homo habilis* and may also give rise to the South African robust australopithecines.

The thought that *habilis* is the direct descendant species of *africanus* is regaining credence after it was pushed out of the family tree by *afarensis* in the mid-seventies. But don't be lulled into assuming that even this sorts out the complexity, because in recent years another question has arisen. Is *habilis* really a member of the genus *Homo* as Louis Leakey, Phillip Tobias, and John Napier suggested back in the early sixties? Or is it in fact an australopithecine? Recently, several researchers working on the *habilis* material from East Africa have suggested that the hominids presently assigned to *Homo habilis* may more appropriately belong in the genus *Australopithecus*. In other words *Homo habilis* is actually *Australopithecus habilis*! Another very plausible option and one that is gaining some credence is that the animal we call *Homo habilis* is just a late version of *Australopithecus africanus*!

So that brings up the question of what criteria are being used to identify a member of the genus *Homo*? We may see in the first years of the 21st century the idea gaining ground that *habilis* should be moved out of the direct ancestry of the genus *Homo,* while a lesser-known species takes its place at the top of the *Homo* lineage. New fossil discoveries suggest that there are no fewer than two other species of hominid classified as early *Homo* wandering around between 2.5 and 2 million years, so there are plenty of candidates if *habilis* should fall by the wayside.

Is stone-tool manufacture, then, a criteria for membership in the genus *Homo*? The answer is no. We've moved beyond the archaic idea that simple toolmaking defines our genus. More and more evidence is suggesting that tool manufacture goes back much further than we thought, now beyond the 2.6-million-year mark. And brain size? Again the answer is no. The cerebral Rubicon of around 600 cubic centimeters for the genus *Homo* has been approached, and surpassed, by early gracile australopithecines. So what are the criteria? The simple answer is that given the variability emerging in the hominid fossil record, we just don't have a universal rule that will allow us to say "this is *Homo* and this is not." Perhaps more and better fossils from the late Pliocene will sort this problem out, but we may be a long way from firm answers.

And what about the newly named species *Australopithecus gahri* recently described by Berhane Asfaw and colleagues. With its enormous post-canine teeth and long forearms (presuming that the partial skeleton found nearby belongs with the cranial material), it's not at all clear where this hominid comes from. It has such a combination of morphologies that it could be descended from either *afarensis* or *africanus,* or something else entirely.

As muddled and challenging as this new information coming from the bone fields of Africa is, I am certain that shortly after this apparent period of diversification in the hominid line around 1.7 million years ago, a single species, *Homo erectus,* emerges as the first hominid, which almost any scientist would agree is definitely a direct human ancestor.

CHAPTER 10

SKELETONS
IN THE CLOSET

~~⌒~~

There are two things most scientists hate dealing with: money and other people. Late in 1996, when Phillip Tobias nominated me as his successor, I reconciled myself to the notion that dealing with these two issues would be an integral part of my job as a scientist. I was excited and a little intimidated by the challenge facing me as director of the Paleontological Research Group at Wits. I knew the job would not be easy. The university's research program was world famous, and I stood as only the third head of paleoanthropology at Wits over a 75-year period in which the two previous incumbents—Raymond Dart and Tobias—were among the most famous scientists of the 20th century.

At the time I took over, the research group had more than 20 employees, a number that would soon grow to more than 50. It has been said the task of managing academics is like herding cats, except that cats are not as independent-minded as scientists. And money? It is one thing to justify funding if the donor's projects are yielding results, but it is very difficult to petition for cash if the perception is that the money is being used to dig holes that yield nothing. Wits was thirsty for cash, with excavations underway at more than nine different sites, including Sterkfontein, Drimolin, and Gladysvale. My most daunting task was to raise the enormous amounts of funds that such a research entity consumes on an annual basis. The part of the job I looked forward to most was the responsibility of curating the university's large fossil hominid collection and, in particular, working through the backlog of undescribed specimens. The trickiest aspect of my appointment would be to work with and direct the research of scientists associated with the group. I knew that in this regard my biggest challenge would be Ron Clarke.

Clarke was the most senior scientist employed by the research group and in charge of the field operations at Sterkfontein. He has had an extraordinary association with the field going back to the days of Louis and Mary Leakey, under whom he had worked as a technician before doing his Ph.D. with Phillip Tobias. His particular expertise was in fossil reconstruction, and there are few people in the world who are better than Clarke at taking fragmented pieces of fossil bone and putting them back together into a meaningful shape. Because he had also applied for the position as director—and at the age of 52 he was 20 years my senior—I expected some degree of tension between us. But we had worked well together on previous projects, and I hoped that the collegial nature of our relationship would continue. I was acutely aware, however, that Clarke was a fiercely independent scientist.

Clarke had come to South Africa from Tanzania following a falling out with Mary Leakey while working as a technician at Laetoli on the famous footprint trail. He had lost his right to enter Tanzania when, without permission, he crossed the border from Kenya into Olduvai Gorge to join an excavation. During this unauthorized foray, he was stopped by the Tanzanian authorities and prohibited from entering the country, a ruling that would stand until the early 1990s. His skill as a great fossil reconstructor became apparent when he worked under Tobias as a Ph.D. candidate on the well-preserved face of SK 847, a *Homo erectus* from Swartkrans. Bob Brain recounts that on numerous occasions when a new find from Swartkrans arrived, he would see Clarke sitting in the hominid safe at the Transvaal Museum working on the specimen. Clarke would eye the specimen for a moment then rush over to a cabinet to find a piece of fossil that would fit onto the new find perfectly, a fossil that may have been unearthed decades earlier.

Ultimately, Clarke would have as stormy a time in South Africa. While working for the Bloemfontein Museum at Florisbad, an Archaic *Homo sapiens* site in the Free State, he had conducted extensive excavations in a large spring mound. After he had accepted an appointment with Wits, he accelerated the pace of the excavation in an apparent attempt to recover as many hominid remains while he still had contractual tenure. In the process, he incurred the wrath of some of the museum authorities who were concerned about the preservation of the context of the site.

Clarke's biggest fallout, however, was with Tobias. In 1985, Wits University hosted the Taung Diamond Jubilee to celebrate the 60th anniversary of the announcement of the Taung Child. Although the academic boycott was then still in force, scientists from around the world came to South Africa to participate in a series of symposia on human origins. Among them were luminaries such as the great British anthropologist Michael Day, Lucy's "parents" Don Johanson and Tim White, and many others who arrived in Johannesburg for this extraordinary event. During the proceedings, Clarke used his platform to launch a personal attack on Mary Leakey, an action that caused an uproar among the delegates. Within a few months, Tobias had Clarke declared persona non grata at Wits, and Clarke and his wife, Kathy, left South Africa for the United States, not to return for seven years. The bad blood from that incident still appears to flow in the Leakey veins. Twelve years after that incident, Mary's daughter-in-law Meave Leakey threatened to withdraw from the Dual Congress at Sun City if Clarke was allowed to speak at the same colloquium she was organizing.

Clarke returned to Wits in 1991 to take over the excavations at Sterkfontein when Alun Hughes, director of the field operation, fell ill and was unable to continue his work. Hughes, a heavy-set Welshman who had run Sterkfontein since the excavations were reopened in 1966, specifically requested that Clarke be brought in to take over. Tobias relented, and the two men made the best of having to get along professionally. It wasn't long, however, before there was renewed friction between Clarke and Tobias.

In 1995, Sterkfontein hit the headlines once again with the discovery of Little Foot, the remains of a foot bone of an as yet unassigned australopithecine. The foot bones had actually been recovered in 1978 by Hughes, but his team had failed to identify them as hominid, and they had been packed away into boxes along with other fossil material from Member 2—one of the older Sterkfontein deposits. In late 1994, Clarke reviewed all material from a large underground chamber known as the Silberberg Grotto, which was part of Member 2. The work was part of a process of examining the backlog of fossils that had been recovered during the Hughes era but which had yet to be properly sorted. While going through a number of boxes, Clarke happened upon an extraordinary find. A bag within one of the boxes contained six hominid foot bones, five belonging to a single left foot and one, a small bone of the foot known as the lateral cuneiform, belonging to a right foot. This set of fossils, quickly dubbed "Little Foot," was important on two counts. First, it was rare to find early hominid foot elements so closely related to each other, and second, the analysis of these bones would provide clues as to how bipedalism developed in southern Africa.

As Tobias was his immediate supervisor, Clarke was obliged to invite him to assist in interpreting and analyzing the foot bones. Their finds were published in *Science* on July 28, 1995, and caused an immediate controversy. The two men argued that

Little Foot was the oldest hominid from South Africa with an estimated age of 3.5 million years. This was disputed by several academics who predictably raised concerns about the unreliability of dating South African fossils. But what really touched a nerve was their claim that Little Foot was prehensile as well as bipedal. Prehensility is basically the ability to grasp, and what Clarke and Tobias maintained was that although Little Foot had a humanlike big toe, it also displayed chimpanzee-like characteristics. They argued that Little Foot, although able to walk upright on the ground, was still comfortable with an arboreal existence.

Many researchers who had viewed the illustrations in *Science*, felt that the morphology of Little Foot was not as primitive as Tobias and Clarke claimed. Therefore numerous requests for casts of the fossils were received by the lab at Wits. I was surprised by Tobias and Clarke's reaction. They held their cards (and fossils) close to their chest, not allowing casts of Little Foot to be supplied to any researchers. Nor were most visitors to the department allowed access. Even Henry McHenry, an old friend of both Clarke and Tobias, was not allowed to examine the specimens. When McHenry came to work with me on the body-proportions research in early 1996, the Little Foot fossils were removed from the fossil safe by Clarke and put into his private lab safe for the duration of his visit.

One scientist who would not take no for an answer was the Great White Shark. When Tim White brought the Dream Team to South Africa, they were determined to compare all the significant South African fossils with their own East African material, including *ramidus*. To have refused their request to look at Little Foot would have provoked the scientific equivalent of a diplomatic incident. Nonetheless, my attempts to show them the Sterkfontein fossils were almost thwarted. Clarke happened to be overseas when the Dream Team arrived in Johannesburg, and the Little Foot bones were still locked in his own safe. It took a great deal of persuasion on my part to get Clarke's wife to come in and open the safe so that the specimens could be examined. The Dream Team was dismissive of Little Foot. After they studied the bones for several hours, comparing them with the hominid casts they had brought with them, White and his group came to their conclusion. Clarke and Tobias, they said, had overinterpreted the primitiveness of the foot bones. Their belief was that Little Foot was largely human in its morphology and exhibited little or no signs of prehensility in the big toe, an assessment that I tended to agree with having examined the fossils on numerous occasions myself. Tobias was unimpressed with their findings and dismissive of the Dream Team.

How is it that two groups of scientists looking at the same set of bones come up with two distinctly different results? The debate centers on two bones of the foot complex: the medial cuneiform and the navicular. In their 1995 paper, Clarke and Tobias discussed several aspects of the Little Foot bones that they felt indicated adaptations to prehensility. But their focus on the mobility of the big toe—their strongest argument—was the morphology of the cuneiform and navicular bones. In an illustration for the *Science* article, Clarke and Tobias had drawn a hatched line to indicate where the contact facet on the medial cuneiform might have been, and used an arrow to indicate the extent of the facet on the navicular. The Dream Team disputed the position of these contact facets as Clarke and Tobias had drawn them, and argued that this had led to an overestimation of the mobility of the joints. In addition, they felt that the basic overall morphology of the foot complex of Little Foot as a whole did not differ substantially from that of the OH 8 foot from Olduvai Gorge, a well-studied specimen attributed to *Homo habilis* and generally considered to be largely humanlike in its morphology. This dispute has never been settled in the literature.

Despite the united front Clarke and Tobias put on for the outside world, Little Foot strained their already sensitive relationship. Clarke told me that Tobias had muscled in on his discovery and had claimed all the credit. Certainly most of the press about the find featured Tobias extensively, partly because he is such an institution in South African paleoanthropology and partly because he is such an engaging showman. The South African *Sunday Times*, for instance, published a photograph of him smiling coyly, with his trouser leg rolled up and his bare left foot on the table next to a reconstruction of Little Foot.

If Clarke was justified in feeling aggrieved that Tobias had sidelined him in the popular press, he only privately expressed this resentment. Nonetheless, his output appeared to suffer following their work on Little Foot, and he did not complete the skull reconstructions that he had undertaken on specimens from Sterkfontein and Drimolen.

Following the announcement in late 1996 that I was to take over the paleoanthropology unit from Tobias at the beginning of 1997, I knew I would have to talk over a variety of issues with Clarke. The two of us met to discuss how we would work together and to review Sterkfontein, which, according to his annual report, was consuming over half the departmental budget but not producing a lot of research papers. As I was now the chief fund-raiser for Sterkfontein, I impressed upon him the need for more productivity. I also wanted to see some changes in the way fossils were excavated and mapped. I felt that the creation of a detailed map using a laser theodolite and moving to a pinpoint mapping system like we were using at the other sites was the only way to go. Clarke disagreed, arguing that we needed to move more earth to find hominids and we didn't have forever. Despite these differences, I renewed his annual contract as senior research officer in PARG, with an agreement that we would both revisit the situation at the end of the year. The renewal of Clarke's contract was based on the understanding that he would be allowed to work at the Sterkfontein site independently, as long as he would keep me fully briefed on all developments. Clarke agreed.

I did not think that Tobias would be an issue. After all, he had retired from the university, was due to hand over his unit to me, and had no formal responsibilities or obligations that would result in his crossing paths with Clarke. Two weeks before I took over the unit, a matter was brought to my attention that should have signaled that something was wrong. One of my first new allies was Tobias's former secretary, who proved invaluable in assisting me in the transition, showing me the ropes, helping me through the mountain of correspondence relating to access to the fossil collections, and to every other aspect of running this historical entity. Just before going on Christmas leave, she called me into the office and shut the door.

"I thought you should see this," she said, handing me a National Monuments Council Permit Application form, which enables an excavation to take place. Only one person may hold a permit. This is usually the person excavating the site, not necessarily the head of the organization that employs them. In the case of our excavations, the permits were always signed by Tobias, as director, and this would be a job that fell to me from January 1. The application in my hand was the renewal of the permit for excavation at Sterkfontein for three years from the day I was to take over. The old permit expired on December 31. Tobias had put himself and Clarke on the permit application as "co-permit holders" and signed it himself as director. I was surprised that he didn't discuss it with me, particularly since technically this application was going to be processed under my leadership. Nevertheless, I wasn't overly concerned.

"So?" I asked.

"I just thought they should have consulted you first. You are, after all, funding the whole operation and will be the boss."

She was right to be concerned, but my focus was on breathing new life into the department. My immediate priority was to tackle the backlog of undescribed material from Sterkfontein that had built up during the previous years. I decided to restructure the unit by assembling a small and dedicated group to work on various aspects of the collection. After mulling the issue over, I decided to include Clarke because of his familiarity with Sterkfontein and his expertise in reconstruction. But because I was concerned about his ability to meet the strict deadlines that I wanted to introduce, I decided to pair him with Charles Lockwood. Lockwood is an extraordinarily competent and hardworking scientist who had just completed his Ph.D. under Tobias on the facial morphology of the Sterkfontein hominids. The two of us got on very well together, and I thought the energy that he would bring to bear in writing up the cranial remains would counter Clarke's recent slowness.

By far the most daunting task was the description of the dental remains that made up in excess of 80 percent of the collection. I decided to assign an Italian scientist named Jacopo Moggi-Cecci. Moggi-Cecci had proved himself reliable and trustworthy during his annual excursions to South Africa during the past decade, when he'd done his postdoctoral studies under Tobias. I would handle the postcranial remains myself and make use of the assistance of several graduate students and colleagues from around the world. All in all, I felt satisfied that I had a workable scenario for getting most of the Sterkfontein fossils published within two years. I was confident I had the right team to do justice to what promised to be one of the most exciting eras of paleontology in South Africa. Once we had this material out in the literature, the scientific world could compare, for the first time, the large South African sample with the other early hominid samples in the rest of Africa.

My focus also expanded to include a new project in the South African province of the Free State, where I had discovered a series of rich fossil deposits in erosional beds in the Sand River. These fossil beds were very much like those in East Africa, formed by either small lake systems or riverine deposition. From the initial surveys, they seemed to be on almost every river system down to the Orange River in central South Africa. The possibility of finding rich surface fossil deposits that we could compare with the cave sites was too much to pass up. This initiative was a collaboration with James Brink at the National Museum at Bloemfontein, Steve Churchill at Duke University, and Peter Unger at the University of Arkansas. This survey work required the development of an exploration team that could identify exposures that needed to be worked on. The idea was, and is, to create an atlas of sites in the first few years of the project so that we could identify just what was available to work on in these systems. A number of new graduate students joined me to work on new finds, including Deano Stynder, who had been on the original Hoedjiespunt dig in the Cape, and Aaron Hutchinson, who would take over an important new dig we had discovered near Bloemfontein.

The one problem that kept nagging at me, however, was Ron Clarke. Andre Keyser, who was in charge of the Drimolen excavation, told me that he was frustrated by the pace of the reconstruction. It was simply not getting done. His normally genial voice crackled with exasperation on the other end of the telephone line. Four years ago, Clarke had promised a "quick reconstruction" of the robust skull from Drimolen but had not delivered anything. Keyser said he'd seen the specimen locked away in Clarke's private safe and that it looked as if little, if any, reconstruction work had been done on it. The skull was the most important part of the Drimolen assemblage, and Clarke was holding up the whole project. Several interesting teeth and postcranial fossils had been found at Drimolen, yet nothing could be published until the skull reconstruction had been completed.

This wasn't the first time this issue had surfaced. Almost as soon as I'd taken over the running of the department, Tobias sent me a copy of a letter that he had written to Clarke raising concerns about his performance dating back three or four years. The two-page memo detailed the problems of the Sterkfontein and Drimolen skulls, and Tobias noted rather cuttingly that Clarke, to the detriment of his employers, seemed to have plenty of time to travel abroad and do freelance work for other institutions. He was referring to two very complex reconstructions that Ron had recently completed—one of a crushed *Homo habilis* skull from Olduvai Gorge and the other a Miocene ape fossil from Italy.

In better days, Phillip Tobias (left), Andre Keyser (middle), and Lee Berger discuss the announcement of the discovery of the Gladysvale hominids.

Additionally, Tobias felt that Clarke's reporting on Sterkfontein was inadequate. But the most damning point in the letter, and one that left a profound impression on me for the rest of the year, was the accusation that Clarke had been inappropriately and impermissibly keeping fossils in his private safe and even at his home. I felt distinctly uneasy at this criticism of Clarke's ethical behavior—the last thing I wanted was to be caught in the middle of the ongoing fight between Tobias, my predecessor, and Clarke.

So it was with some concern that I listened to Keyser's complaints that the slow pace of progress could impact on the funding he needed to continue excavating Drimolen, which was beginning to fulfill its promise of being a rich fossil treasure chest.

I decided to confront Clarke directly on the issues raised by Keyser and Tobias. We met in my office, where I explained that in order for us to have a good working relationship, he needed to address the issues raised in the Tobias memo and improve his productivity. I pointed out that the policy of the university was quite clear about the storage of fossils and that any material he might be holding in his private capacity needed to come back to the hominid vault immediately. Clarke did not argue. He admitted that he had a few skulls in his own safe but that these were for the purposes of reconstruction, which he said he would resume immediately. I was partly reassured by his attitude but decided to keep a close watch on him over the next few months.

I was determined not to let the Clarke issue get in the way of my overall management of Sterkfontein. Even though it is the world's longest-running paleontological dig—having been continually excavated since 1966—there was still a lot more fossil material that could be recovered and fundamentally affect our view of early human origins. The site is a magnificent geological history book, and although its main value has been in the fossils preserved over the past 3.5 million years, Sterkfontein is almost as old as time itself. Once an ancient seabed, the valley contains some of the oldest undistorted rocks on the planet, dating back 2.8 to 2.6 billion years. That's just over half the age of the Earth itself. These dolomites, comprised largely of calcium carbonate, manganese, and silica, are the sedimentary remains of an ocean floor that hosted some of the earliest forms of life on Earth: The fossilized remnants of prehistoric blue-green algae can still be seen within the rocks today.

When Tobias first opened that safe for me in 1989, there were no fewer than 550 cataloged specimens from Sterkfontein alone. This tremendous assemblage was almost wholly due to his and Hughes's efforts. The two men had been collecting fossils almost continually during excavations at Sterkfontein since 1966, when Wits University rescued the site from the official neglect it suffered in the wake of the heady days of discovery during the Broom era in the thirties and forties. The apartheid government, uncomfortable philosophically with the anti-creationist nature of paleoanthropology, would have been happy to let the Sterkfontein drift into obscurity had Wits not intervened. Since that time, Sterkfontein has still had more than its fair share of uncertainty. It has faced closure because of a lack of funding, and there have often been lengthy periods of little yield. The renewed excavations, under the leadership of Phillip Tobias, started slowly. Hominid fossils in situ were disturbingly absent, and the sheer labor of removing the tons and tons of lime rubble occupied the university team for almost a decade. Over 50 good specimens were recovered during this process, found in loose breccia blocks, so their context within the site is largely unknown. They are really only positioned by either their morphology or preservation.

Alun Hughes was the driving force behind Sterkfontein after its rescue by Wits University. Noted for his organizational skills, he acted variously as chief technician of the department of anatomy and as field researcher. He was responsible for daily operations at Sterkfontein, where he'd spend two to three days a week at the site supervising the eight excavators and preparators. It was into his hands that most of the fossils would fall. And it is in his tight, neat handwriting that he records in black ink the catalog number and location of almost every fossil in the safe. Hughes was a contemporary of the great James Kitching, who found the ape-man skullcap at Makapansgat before going on to establish himself as a Karoo fossil finder of note. But he had also been a fixture in the South African paleontological community since the days of Dart, for whom he'd dig up hyena lairs to disprove claims that they were bone collectors.

I had the privilege of working with Hughes in 1990, two years before he died. The heavyset Welshman taught me more about the dolomitic caves and excavation than I could have learned from any textbook. But as is often the case in this profession, Hughes was also in some ways embittered by the treatment he had received from the institution for which he worked. He had never forgiven the university for not awarding him an honorary doctorate. Instead, he was given a Masters *honoris causa*, an award that had never been given at the university before, and one that he felt fell short of recognizing his contribution.

During the 1970s, our understanding of Sterkfontein took a major step forward when the site geologist, Tim Partridge, tackled the complicated stratigraphy of the deposits. He labeled the sequence of jumbled sediments Members, starting with the oldest, Member 1, and dating back beyond the three-million-year mark to Member 6, which, at a mere 100,000 to 200,000 years old, is the youngest. The richest deposit so far has been Member 4, which contains most of the *africanus* specimens used in arguing the long arms–short legs hypothesis.

It was not until 1976 that the first good hominid was recovered from a context that could clearly be identified. That was a tumultuous year for South Africa. Black revolt against white rule flared up in what was known as the Soweto riots. As was the trend during the seventies and eighties, any finds in South Africa were overshadowed by the country's politics. The specimen, Stw 53, remains one of Sterkfontein's finest. At the time, it was announced as the first *Homo habilis* specimen from South Africa because of its less protruding face, head shape, tooth-wear pattern, and a brain capacity that was perceived to have been greater than that of the ape-men. It was assigned to Member 5, a bed where early Acheulean and Olduwan stone tools had been discovered. Stw 53 was later reconstructed by Ron Clarke, who suggested that it might be *H. ergaster*. This specimen remains in contention today, with some scientists suggesting that it is in fact a late *africanus* specimen, representing a transition toward early

Homo. The final word is still out on this specimen, but almost everyone now looking at the cranial specimens agrees that what Stw 53 is not is *Homo habilis*.

Over the following decades, the Sterkfontein excavations moved from west to east across the site—becoming richer and richer in hominid fossils. In the early eighties, Clarke, then a Ph.D. student, began his association with Sterkfontein by assisting Hughes and Tobias in excavations. By the late 1980s, excavations at Sterkfontein were penetrating down into a partially decalcified area that has often been referred to as the Swallow Hole, due to its abundance of australopithecine fossils. It is from this cavity, which measures only about eight yards across and ten yards deep, that most of the specimens in the university's hominid safe have come. Almost all of the specimens are relatively well pre-served with very little crushing. But one of the problems is the context of the finds. A grid system was originally established at Sterkfontein in the 1960s, consisting of steel girders, wire, and nylon fishing line. The excavation was conducted in spits that measured one yard by one foot. By 1960s's standards this was acceptable, but by the mid-eighties the necessity for greater accu-racy was recognized around the world.

Perhaps due to South Africa's political isolation, however, the excavations at Sterkfontein continued at this relatively crude level. Years of temperature fluctuations between the cold high-veld winters and blazing summers made this demarcation even rougher, and over the years the grid had warped slightly in sev-eral directions. If that wasn't enough to drive precision fetishists mad, much of the excavation was done with laborers using picks and shovels. Material was poured into buckets and carried away to be sieved and then later sorted and identified. This meant that the most accurate positioning of any single fossil was one square yard by one foot or within nine cubic feet.

This led to all sorts of problems of association. Back in the lab in Johannesburg, one could never determine the original positioning of two or more fossils from the same or nearby spits. In an area as small as the Swallow Hole, with its hundreds of fossil hominids, this has often proved to be disastrous. Often two or more clearly separate specimens appear mixed together in the assemblage. For this reason, I was determined to modernize the way excavations at Sterkfontein were carried out. The site still used the wire-framed grid system, which I felt should be replaced with a more modern laser-mapping system, like the ones in use at Drimolin, Gladysvale, and Hoedjiespunt. In addition, the Sterkfontein technicians still largely used only hammer and chisel to remove fossils from the breccia, pretty much the same tools used by Robert Broom and Raymond Dart. I felt that new methods should be explored for extracting these fossils; I wanted to give some of the newer techniques, such as acid preparation, a chance at the site.

Clarke's reluctance to entertain any change to the way Sterkfontein was run became increasingly frustrating. In addition, no fossils appeared to be coming out of the rocks. There was no indication from Clarke that he had any publications in the pipeline. Another source of frustration was the fact that there were four more skulls from Sterkfontein Member 4 that needed reconstructing before the descriptions could be published, and none of these had been worked on.

My trips to Sterkfontein weren't providing me with much information. I rarely found Clarke there. The men were not excavating but were preparing isolated blocks of breccia. The atmosphere at the site was distinctly odd. The men, who were usually congenial, were quiet and tried to avoid answering any questions. In retrospect, I should have been suspicious, but I had a lot of other things on my mind.

One was the incredible variation among the specimens of *africanus* that had been recovered from Sterkfontein. To understand the proposal that *africanus* evolved not just into one creature but possibly two, one has to look more closely at this variation.

The jumble of bones in South African caves makes it difficult to temporally associate them, but over the years our research there has come up with an estimation that all the bones discovered so far from this site represent about 120 *africanus* individuals. These skeletons were far from uniform in their morphology, a case in point being Stw 505, a male *africanus* skull that Hughes had found in 1989. Although this had all the characteristics of a gracile australopithecine, it's brain capacity was far bigger than any of the specimens we had found to date. In fact, with an internal skull measurement of between 620 and 630 cubic centimeters, it had crossed the rather arbitrary cerebral rubicon of 600 cubic centimeters that is accepted by many as the entry-level brain size considered for membership in the genus *Homo*. Either Stw 505 is challenging us to rework our definitions or the australopithecines at Sterkfontein were beginning to show a remarkable increase in brain size while still maintaining their primitive morphology.

At the time, this was considered to be a surprising result, but for some reason Tobias did not want to publish this hominid fossil; it sat in the safe for years. Eventually, Clarke gave a photo of the specimen to Don Johanson for his book *From Lucy to Language* and passed on the information that the brain capacity was about 625 cubic centimeters. This prompted a debate about whether the reconstruction done by Hughes was correct. Glen Conroy, Tobias, and others prompted a debate in the journal *Science* by saying that CAT scans of Stw 505 showed that the skull only had a capacity of around 515 cubic centimeters. Most of the researchers who had worked closely with the 505 specimen in South Africa, myself included, feel that Conroy's team did not sufficiently take into account the distortion of this skull, but the matter is still to be settled.

During the remainder of 1997, I tried to get Clarke to be more proactive in analyzing the several hundred new hominid fossils in the Wits hominid vault, as well as in cataloging the several hundred thousand animal fossils that had emerged from Sterkfontein over the previous three decades.

Everything else at Wits seemed to be going exceptionally well. I was confident that 1998 would be just as good. By early January we were in full swing preparing for the July Dual Congress. The survey projects were underway in the Free State, and we had discovered a fantastic new site near Bloemfontein called Erfkroon (Inherited Crown), one of the most extensive surface assemblages ever discovered in southern Africa. I had received two honorary adjunct professorships, one from Duke University in Durham, North Carolina, and one from the University of Arkansas. In our first field season at Hoedjiespunt, we found a tibial shaft of a human child that was unexpectedly robust and was our first good piece of postcranial bone.

Then the issue with Ron Clarke came to a head. While he was away on an overseas trip, we discovered that we needed a fossil he had removed from the hominid vault. I was sure that he would have kept it in his private safe, so I called his wife to bring the key to the department. I was astonished to see that the safe contained literally dozens of hominid fossils that I had never seen before. I clearly saw a femur, several finger bones, skull fragments, and a hominid clavicle. I was furious, but I knew I couldn't take my anger out on Clarke's wife. Instead, I retrieved the fossil that I'd requested and walked out determined to resolve why he was continuing to flout the university's regulations by keeping fossils outside the main safe. I checked the university's main fossil catalog and confirmed that there were no new entries since Little Foot, which meant that Clarke was the only person who knew about these fossils. Coincidentally, I received a second complaint from Keyser that the Drimolen skull remained in pieces.

The situation had become untenable. I confronted Clarke on the fossils in his personal safe and he duly returned the specimens to the main safe and entered them in the catalog. One such bone was a clavicle he had kept from the department for four years. The only restitution available, I believed, was not to renew his contract. I sought Tobias's advice on this course of action and he assured me of his support. I duly drove to Sterkfontein in February 1998 to tell Clarke that his contract was not going to be renewed. He took the news with a philosophical shrug of the shoulders, saying, "So be it."

My sense of relief as I left that meeting was tinged with regret. Clarke had left me no option but to refuse renewal of his contract. Yet he had been an integral part of Sterkfontein and, despite his recent low productivity, had made a valuable contribution, particularly in alerting us to the possibility that the Sterkfontein *africanus* collection may contain two species, not just an anatomically varied single species. This variation may be due to a number of factors. On the one hand, *africanus* may just be a highly variable species with a great degree of differences between individuals. Or it could be that the Member 4 area, where most specimens have been found, is not as homogenous as previously believed—that is, it represents different layers of time and ecologies mixed together. The third option is that there is more than one species.

During the eighties, Clarke had published his description of the Stw 252 skull, a juvenile male that had very large teeth and the kind of flat face that suggested it was closer to a robust than a gracile ape-man. He believed that this was in fact the ancestral form of *robustus*, and he later identified a number of other specimens, including Stw 505 as part of this "proto-robust" group.

Charles Lockwood, who has done the most comprehensive and detailed study of the crania from Sterkfontein for his Ph.D., thinks Clarke is both right and wrong. In some cases, he thinks he is right for the wrong reasons. Lockwood agrees that there is probably another rare, more robust species than *africanus* in the collection. He differs substantially, however, on which of these specimens they are. His work, which is yet to be published in detail, recommends that we tread around the cranial remains of

Member 4 at Sterkfontein and Makapansgat very cautiously. His study of the fossils from both these sites indicates that there may have been an early divergence of *africanus* leading to a robust lineage before three million years, and a later divergence toward *Homo habilis* around 2.8 million years.

After the Dual Congress at Sun City, I learned that Clarke had been offered a job by a German group. I was glad; the decision not to renew Ron's contract had been a tough one for me emotionally. I didn't warm to this kind of power confrontation, and it would have tugged at my conscience had he not found another job once his Wits contract expired. By the end of July, the department settled back into its regular routine in the labs, and we got down to working on the descriptions of the material. In particular, I started focusing on getting a monograph out on the 431 skeleton that Tim White had given me such a hard time about.

At about this time Fred Grine, a former Ph.D. student of Tobias's and now a professor at the State University of New York at Stony Brook had a chat with me about the fossils. He told me a story that astounded me. In the early 1980s, he had been approached by Tobias to describe the dentition of the specimen from Sterkfontein. During their discussions, Tobias invited him to make a broader study of the Sterkfontein dental collection, which comprised several hundred individual teeth. Fred had agreed and had duly described all the specimens, handing in a finished manuscript to Tobias. But he and Tobias had fallen out shortly thereafter, and Tobias and Grine parted ways, leaving Grine with valuable material that he was not allowed to publish. I was appalled and immediately discussed the issue with Jacopo, who was well into the tooth descriptions. He agreed that it would be a waste of scientific energy not to incorporate Grine's work. We agreed that this would not only speed things up but also right a past injustice. Grine was thrilled and the preparation for the descriptions of the teeth was given a big boost.

Most of the specimens locked up in the hominid vault at Wits University have been attributed to *africanus*, probably the most represented species in the pre-two-million-year-old African fossil record, with only its old rival *afarensis* having close to as many elements available for study. One of the problems facing the new team was how to work out which bones belonged to which individual skeletons. This is a complicated exercise, particularly when a single species dominates so much of our fossil record. The answer is in a process called MNI (the Minimum Number of Individuals at any one site).

At last count there were just over a thousand cataloged specimens of *africanus* in the collections of my research group and those of the Transvaal Museum. The vast majority of these come from just one site, Sterkfontein. The remainder, maybe 50 specimens, come from Makapansgat, Taung, Gladysvale, and Bolts farm (the latter two sites being in the Sterkfontein vicinity but yielding only a few isolated bones and teeth). The thousand or so specimens that make up the Sterkfontein collection can be divided into associated groups—for instance, those that collectively form part of a skull—and isolated specimens that don't really fit in with any of the other fossils.

Surprisingly, most of the Sterkfontein specimens can be associated with other fossils in the collection, which enables researchers to work out the MNI found at the site and gives a reasonably accurate idea of the lowest probable number of living individuals a sample of bones represents. This is done through a combination of methods, including simply counting the number of the most common elements in the collection, say a molar or pre-molar. A more precise method is to assess which bones and teeth might represent one individual and then "minimize" the sample to the lowest possible number of individual animals.

Jacopo Moggi-Cecci and Travis Pickering recently calculated the MNI for Sterkfontein, and came up with a figure of around 100 individuals. This means that out of the 1,000 or so cataloged hominid fossils, we have a lot of bones and teeth from a few individuals at Sterkfontein. This is probably due to the nature of the cave deposits themselves and the way in which the bones find their way into the caves. So taphonomy is a useful tool to try to fuse the blessings—in the form of excellent fossil preservation—and the bane—geological complexity—found in South African dolomitic cave sites. Nonetheless, absolute dates for these fossils are difficult, if not impossible, to achieve.

This work enabled us to start coming to grips with the "speciation window" that we suspect occurred at Sterkfontein between 2.8 million and 2.6 million years ago. In August and September the two scientific papers and my NATIONAL GEOGRAPHIC article on long arms–short legs were published. Attention was focused once again on the South African hominids. I was pleased in general with the responses to the work. It seemed like most scientists grasped the methods and significance of the results and were ready to accept a completely different animal in *africanus* than the one that had existed in their collective perceptions. I was surprised by the silence emanating from Tim White's Berkeley group, given that we were attacking the position of their beloved *afarensis* in the human family. The reasons for their silence would emerge the following year.

In August 1998, I received a letter from Tobias that indicated he had changed his attitude toward Clarke. Tobias said that Clarke had been a brilliant scientist and an excellent researcher and insisted that I rehire him immediately. This was completely contrary to the advice he had given me barely seven months before, and yet he provided no explanation. The letter was very unlike the Phillip Tobias I knew. I responded, outlining the background to the situation and pointing out that after certain scientists wrote to the deputy vice chancellor requesting me to review my decision, I actually had offered Clarke his position back again subject to certain conditions. He had refused.

Tobias wrote back a short note saying he didn't want to go into details as to why he disagreed with me but felt I should take his advice and rehire Clarke. This letter was uncharacteristic of Tobias, and it was such a vehement shift from the position that he had long held and from the support he'd given me to date. I was so alarmed by this reversal that I took these letters to the vice chancellor of Wits, Colin Bundy, who believed that I had acted appropriately and that I was justified in standing by my judgment in this situation.

Barely a month later, on September 25, Ron Clarke called me at home and said he had something to show me. I felt distinctly uneasy. I hadn't had any further communication from Tobias, but his letters had rattled me. I felt there was something going on that I wasn't party to. Clarke refused to give me details over the phone. He wanted me to come to Sterkfontein. I was due to be there on Monday, September 28, for a meeting to discuss the application for the Sterkfontein Valley area to be declared a World Heritage Site, so we agreed to meet on-site afterward. It was with some trepidation that I drove out through the Krugersdorp hills to Sterkfontein that Monday morning, wondering what lay in store for me

Tobias and I, accompanied by the head of the Wits anatomy department, Beverley Kramer, met Clarke at the research hut alongside the excavation area.

"I've got something really interesting to show you." He smiled with a slightly strained nonchalance, refusing to be drawn into divulging any more details. Skirting the public entrance and the bust of Robert Broom that stands guard there, we followed Clarke through the back entry and down into the underground cave system. The eerie echo of our footsteps accompanied us as we trudged through the darkness of the cavern toward the Member 2 section of the cave.

I could hear some activity ahead of me as we approached, but was unprepared for the blinding glare of lights that struck us as we rounded a rocky outcrop to enter the small chamber. We were being filmed. Shielding my eyes, I noticed the chamber was filled with five people, including a cameraman and sound recorder. Had we walked onto a movie set or the site of a fossil discovery? The answer soon became apparent. Both. Among the silhouettes gathered in the cave, I recognized Paul Myburgh, a freelance filmmaker who'd been intermittently working on a documentary on human evolution in South Africa, and somewhere behind him I could see two of the Sterkfontein excavators, Stephen Motsumi and Nkwane Molefe. What the hell was going on down here? Just then the camera lights passed across us and settled on the cave wall. There in the cold rock, illuminated by the warm swathe of electric light, was an astounding sight: the legs and partial skull of a hominid skeleton emerging from the cave wall.

Clarke, clearly spurred by our stunned reaction, outlined the story behind the find, his voice amplified by the confines of the cave. All the while Myburgh's camera rolled, panning between the skeleton and us. To my astonishment, Clarke revealed that he had found the skeleton in June 1997, nearly 17 months previously, and had kept it a secret. He had been working on the skeleton for nine months before I told him I was not going to renew his contract because of his lack of productivity. Why hadn't he told me about the find? My overwhelming sense of disbelief was tempered by the exhilarating thrill that any new discovery sparks in the heart of a fossil finder. Despite feeling manipulated by Clarke, I found his discovery remarkable. The fossil was clearly going to be interesting. While only the skull, arm, and two halves of the lower legs were exposed, it was clear that it had strange proportions. I looked it over carefully and realized that here was a long arms–short legs test case. The skeleton we were looking at, Clarke told us triumphantly, was the rest of the body of Little Foot, the hominid foot bones that had been found years earlier.

During a sorting operation in May 1997, Clarke came across several boxes labeled D18, which denotes Dump 18 in Member 2. He opened one of the plastic bags inside the box and instantly recognized a small bone on top of the pile as being part of a hominid foot, an intermediate cuneiform preserved in the same chalky white color of the Little Foot bones. As in the case of the first Little Foot discovery, the original excavators of the material had failed to recognize some significant fossils. Clarke then cast his net wider. Over the next few days, he found four more pieces of the hominid's foot in other bags, including a critical fragment of the end of a tibia. He then decided to look through other material accumulated from Member 2. Several days later back at Sterkfontein, on May 27, he found another hominid tibia fragment in a box labeled Dump 20, as well as a damaged calcaneum, the heel bone. When assembled with Little Foot, this collection of 12 bones formed part of the lower legs and feet of a left and right foot that almost certainly belonged to a single individual.

By the end of June 1997, Clarke had secretly cast the specimens and went back to Sterkfontein with a reproduction of the broken tibia fragment. He gave the cast to two of the fossil preparators at Sterkfontein, Nkwane Molefe and Stephen Motsumi, and asked them to descend into the Silberberg Grotto to find a match for the broken end of the tibia with a cross section of bone in the wall of the cave. Now, to appreciate the subsequent events, one has to visualize the Silberberg Grotto. It is a large, dark cave with an enclosed area nearly the size of a four-bedroom house. Its roof lies some 10 to 15 yards below the ground, and one has to descend into the darkness by way of a wooden miner's ladder. Once in the grotto, the surfaces are all damp and covered with mud, so the mere act of just seeing fossil fragments in the wall is difficult. To lengthen the odds against any kind of discovery, Molefe and Motsumi had only handheld lamps with which to search. Nevertheless, they found the rest of the fossil almost immediately. On the second day of the search, July

3, 1997, they came across an exact match for the end of the right tibia in the side of a long slope of hard breccia. The ultimate search for a needle in a haystack had paid off. Even more amazingly, in the rock just to the left of this piece of bone could be seen the broken shaft of the left tibia.

I wondered how Clarke had managed to contain himself and not alert any of his colleagues to this discovery. He did, however, share his confidences with Myburgh who filmed the excavation endeavors on a regular basis. Molefe, Motsumi, and Clarke then began to work on the concrete-like breccia, using hammers and chisels to follow these limb bones upward. All were forbidden to speak to anyone, particularly me, despite the fact that I was the group leader and person ultimately in charge of the excavations. The work was painstakingly slow but initially rewarding. More and more of the lower limbs were revealed and the excavations continued through 1997 into 1998, by which time they'd reached the lower part of the thighbones of two legs lying side by side.

In hindsight, I realize that my decision not to renew Clarke's contract must have prompted him to intensify his efforts to uncover what by then was clearly a complete skeleton lying in the rock. By May 1998, the three-man team had uncovered a whole left radius, one of the long bones of the forearm. Agonizingly for them, just above the broken femurs and radius, they found only limestone and dolomite. As they continued to chip away, they became more and more frustrated. The expected skeleton, which should have been lying in an anatomical position on its back, failed to materialize upslope from the legs. How could the feet and half the legs and one piece of the arm of a hominid be there and the rest missing? They expanded their search for the rest of the skeleton in the cold rocks farther away from the discovery.

For three months the three tinkered away at hard rock with no success. Finally, Clarke took a closer look at the excavation and realized that a fault ran through the cave wall. He worked out that the upper part of the skeleton had possibly been displaced downward and to the left of the legs. He then asked Motsumi to chip away at the breccia in this particular area, and on September 11, Motsumi made a direct hit with the chisel on the end of a hominid humerus. Just above this bone was another fragment of fossil, which, when cleaned, turned out to be part of a mandible that proved to be in occlusion with a whole skull.

At this point, with just three months of his contract left, Clarke had decided to reveal his find and invite Tobias and me to Sterkfontein. I was amazed and exhilarated by the nature of the discovery, but also disturbed that Clarke had been so furtive. I offered him a ride back to Johannesburg so that we could discuss the find. We got in the car together and drove past the Krugersdorp hills toward the distant city lights.

"Ron, why didn't you tell me?"

"I have my own reasons," he replied sulkily.

"Why didn't you tell Tobias, then?"

"Look what happened with Little Foot. I found it and yet Tobias claimed all the credit."

Though Clarke was clearly out of line in keeping the skeleton secret from the university, I was determined that we had to put up a united front and not let the politics of the find compromise the way it was presented to the outside world. Yet right away Clarke and I differed fundamentally on what should happen. He wanted to announce the find immediately. I felt strongly that we should excavate the rest of the skeleton, find out what the damn thing was, and only then make a public statement. Clarke saw this as an attempt on my part to wrest his skeleton away from him. Our discussions ended inconclusively. Unfortunately, the revelation of the find coincided with a prearranged trip I had organized to Botswana.

As desperately as I wanted to resolve the problems around Sterkfontein, I could not postpone the two-week trip, which was the launch of a three-year international fossil survey. I was to survey the exposures of a giant rift lake that the De Beers Diamond Company had discovered during its drilling operations. The lake was buried deep beneath the Kalahari sands, but there was a chance of finding surface sediments due to faulting. The small team going into the bush comprised two graduate students, Deano Stynder and Lloyd Roussouw, and my colleague from the University of Arkansas, Peter Unger. We would be meeting our Botswanan colleague, Nick Walker, in Maun to head into some of the most remote areas in southern Africa in search of fossil exposures. I would be completely out of contact for two weeks.

Before I left, I made a last attempt to convince Clarke that the immediate announcement of the fossil find would be ill advised. I tried to persuade him that the specimen was too important to rush into a media circus without a full understanding of what it was. The 3.3-million-year age would have made it one of the earliest australopithecines found in Africa. It might even be a new species or even a new genus. Clarke's response was aggressive.

"I knew this would happen," he said angrily. "It's just like Little Foot. You are going to try and take the credit for my fossil find."

"Not at all," I tried to reassure him. "You've done a tremendous job. You are still a senior research officer in this research group, and even if you are not going to be with us next year, you will always be a part of this discovery and in charge of the research and analysis of this fossil. It will not be a repeat of Little Foot."

His rank displeasure was obvious and the conversation ended without resolution. I expressed the same opinion to Tobias, who suggested we discuss the matter when I returned from Botswana.

My sense of trepidation as we packed up the Land Rovers and headed west for the desert proved to be justified. While I agonized over what to do about Ron Clarke and marveled at the starlit skies somewhere south of the Makgadikgadi Salt Pans in central Botswana, Tobias engineered a coup. I returned to the university to find that Tobias had seized effective control of the find and that he and Clarke were determined not to surrender the fossil. The permit that Tobias had issued himself clearly stated that he and Clarke were the authentic representatives on the dig.

I walked into a minefield back at Wits. Tobias had taken me completely by surprise. He had persuaded the university to allow him to regain control of the skeleton in my absence. I had assumed that he supported my position, having nominated me as his successor, and that he was my mentor and friend. I was mistaken. I also wondered whether he had been told of the skeleton before we had seen it together, and that his request for me to rehire Clarke was a consequence of this. But this was now immaterial. Tobias had played his hand and obviously saw me as a threat to his continued involvement with the Sterkfontein fossil. Having been an institution at Wits University for almost 60 years, holding the positions of head of the department of anatomy, head of the research group, and dean of the faculty of medicine, no one knew better than Phillip Tobias how to navigate the system. He had initiated a university inquiry into allegedly unethical behavior on my part in purportedly attempting to take credit for the find and other discoveries. I was flabbergasted, and although my name was eventually cleared, I was forced onto a defensive footing, which distracted me from the real issues and sapped my energy. My fortunes had suffered the equivalent of a magnetic polar reversal.

Tobias moved quickly to consolidate his position. While I was wondering how to deal with the situation, I received an invitation—on my research group's letterhead—to attend a press conference to be held in two days time to announce the Sterkfontein find. I then discovered that Tobias had already made a presentation of the find to the Cabinet Ministers and had persuaded the then deputy president, Thabo Mbeki, to participate in the announcement.

On December 9, 1998, Tobias and Clarke presented the skeleton to the world. Despite the fact that the embargo was broken the night before by Communications Minister Jay Naidoo, who bragged about the find at a banquet in India, the press conference was a success, attracting widespread media attention. Tobias announced that the find was "the single most important paleoanthropological discovery of the 20th century" and "the most complete skeleton of an early hominid ever discovered." I was concerned that his enthusiastic speculation to the attentive media was setting the fossil up for more than it could deliver. While the 3.3-million-year-old antiquity of the find—cataloged as Stw 573—is surely spectacular, we have no idea of what species this hominid represents, or what will actually be so significant about it. Nevertheless, the media was enthusiastic about the fossil and Mbeki's presence gave the find a credibility it might not have otherwise enjoyed. Headlines around the world proclaimed South Africa's return to center stage in the search for human origins.

Behind the scenes the relationship between Tobias, Clarke, and me deteriorated. Barely days after the announcement, the tensions between us were leaked to the *Mail* and *Guardian* newspapers, which proclaimed that Wits University had axed its top fossil finder, putting the blame squarely on my shoulders. Clarke was portrayed as this selfless hero who had found the skeleton, and I was cast as the heartless boss who had given him the chop. Tobias came out on Clarke's side and was strongly critical of me, despite the fact that he initially had supported my actions. Nowhere in the story was it mentioned that Clarke had kept the find a secret, and that at the time the decision was made not to renew the contract, he had withheld crucial information from his superiors. Nonetheless, on the advice of Vice Chancellor Colin Bundy, who assured me that he recognized that I had received unjust treatment at the hands of the media, I decided not to comment publicly. For the next few days, I rode out the media storm of unfavorable publicity and weathered the indignity of being named "Idiot of the Week" by the *Sunday Times*.

The University found itself in a quandary in the wake of Tobias's coup. On the one hand, there was the legendary Phillip Tobias and the now famous Ron Clarke holding on tightly to the skeleton and refusing to relinquish it, even though it was the university's property, and neither were employed by the institution any longer. On the other hand they had me, representing their authority over Sterkfontein and playing the role of chief fund-raiser for their excavations. Eventually, through the formation of an advisory committee a compromise was worked out in which it was agreed that Sterkfontein would be ring-fenced as a separate research entity to be run by Tobias, and I was given a promotion to head a new unit to be formed under my name to focus on broader exploration and research. I suggested that the new entity be amalgamated with the Bernard Price Institute, the traditional center of paleontology at Wits, and that paleoanthropology should move off the premises of the medical school, relocating on the main campus. This was passed by Faculty and Senate, thereby resolving one of the most divisive disputes the university has ever faced. My concern, though, remains that this resolution is of a temporary nature and that the university has not resolved the issue by setting up two competing centers of paleoanthropological research.

Now, out of politics and back to the science.

Why is Stw 573 so important? Well, at the time of writing, the skeleton is still in the rock and we don't know if the rest of the bones, the rib cage, spinal column, and other arm, pelvis, and proximal femurs will surface. They are probably there, but only further excavation will tell. We do know, however, that this is the first time that a complete skull of an early hominid has been found in association with good postcranial evidence. This means that a direct correlation can be explored between skull morphology and limb length. Although I've only seen the skeleton once and, of course, only in profile, the skull is clearly that of a gracile australopithecine of sorts. Is it *Australopithecus africanus* or something new? Time will tell, hopefully sooner than later. What we do know is that the sediments in which it has been found have been preliminarily dated to between 3.2 and 3.4 million years old. If these dates hold, this makes it the oldest hominid from South Africa and clearly contemporaneous with *afarensis*. Unfortunately, the excavation and analysis of the specimen still have a long way to go; it will be years before the mysteries of the fantastic Stw 573 skeleton will be revealed. At that point, Henry McHenry and I finally have a good test case for our long arms–short legs theory.

Stw 573, while still trapped in a cold, dark rock face, has played the role of an ambassador for the country. The find provided the university with additional justification for declaring the entire Sterkfontein Valley a World Heritage Site by the United Nations Economic Cultural and Science Organization. Despite an attempt by a casino development consortium to have the UNESCO listing put aside, the Sterkfontein Valley was declared South Africa's first World Heritage Site in December 1999. The listing will ensure that the sites of Sterkfontein, Swartkrans, Kromdraai, Coopers, Drimolen, and Gladysvale are protected against excessive commercial encroachment and that this particular cradle of South Africa's paleontological heritage will be secured into the new millennium.

CHAPTER 11

THE ROBUST
ENIGMA

⤳

The early morning sunlight filtered through the treetops, casting speckled shadows on the leafy floor of the gallery woodland. Towering white stinkwood trees, date palms, and hanging liana vines framed the gurgling streams that laced the network of wooded land between the dolomitic hills of the ancient Sterkfontein Valley. The creeping dawn heralded a cacophony of birdcalls and feathered freneticism in the trees, prompting a lazy stirring in a giant nest wedged in the fork of a fig tree 15 feet above the ground. With slow deliberation, the occupant of the nest blinked, hauled himself into a semi-upright position, and surveyed the terrain around him. In close proximity were several other large nests occupied by the waking forms of his companions. He was momentarily startled by the sudden sharp howl of one of the juveniles, but relaxed when he saw that it was merely a young child's response to the unfamiliarity of a new day.

His community had been badly affected by the leopard attack several days before, when one of the younger females had been dragged off and devoured. For several nights the group had huddled together in their tree sanctuary, unable to sleep, comforting each other in the darkness with consoling grunts. But the daily routine of food gathering had restored some equilibrium to the group, and the intensity of the tragedy had dissipated. Satisfied that all seemed in order, he stretched his long hairy arms, scratched himself, and yawned languorously, displaying a row of huge teeth beneath his broad black lips. His next thoughts were about food. Hunger and fear were the primary conditions that informed his emotional range, and by extension, those around him. And in the morning he was always hungry.

He shinned halfway down the fig trunk, instinctively looked around for any signs of danger, and then dropped to the ground, where he pulled himself to his full height. He was large in relation to the other males in the group, weighing about 176 pounds and standing about five feet tall. His head appeared disproportionately big compared to his body, mostly because his broad, flattish face was flanked by massive jaws. The top of his skull was prominently ridged from front to back, and although this sagittal crest's anatomical function was to anchor his jaws to his skull, it gave him a fierce disposition. He used this to his advantage, having established himself as the alpha male through an alliance with his brothers and his mother, who was one of the oldest of the females in the group. He had perfected the art of looking fierce, using facial threats and expressions to ward off any challenges to his leadership, which entitled him to his harem.

While waiting for the rest of his group of about 20 individuals, mostly females and their young, to assemble for the day's food gathering, he and his two younger brothers wandered toward the nearby stream for a drink. Along the way, they kneeled down and turned over any logs they came across, hoping to find a scorpion or other tasty insect snack. He was obsessive about food, and not very particular about where it came from. His group was sustained mostly on fruit, nuts, and leaves in the wet summer months when food was plentiful, but would not pass by a rat or a small monkey if they managed to catch one. As they walked through the trees, they kept their eyes open for birds' nests in the lower branches of the trees, hoping for an egg or a chick. They also kept a wary eye out for predators, especially the leopards that occupied the caves higher up the valley in the open grassland. One never knew when a leopard might come down to the forest to feed.

In a leisurely way his group assembled, and once he'd noted that everyone was there, he led them slowly through the trees to the more open woodland farther down the stream. Progress was slow as the group moved through the woodland, stopping every few paces to chew on leaves or twigs, grinding the nutrition out of them with their huge molars, chewing continuously on whatever they could put in their mouths. The three adult males kept to the fringes of the group, protectively guarding the females and the juveniles, who stayed close to their mothers. Once they reached the more open country, he planned to leave the group under the guardianship of his youngest brother while he and the second oldest would try their hand at finding a carnivore kill that they could scavenge. The group shared treats like this collectively; it created a close bond. Getting carcasses that were in a sufficient state of dismemberment was a problem. Often, they only got the chance to pick up a scrap or two of bone bearing meat left over by the humans, who, until they were satiated, would dominate a kill. Nevertheless, the more open the country the better the opportunity for scavenging, and meat was meat, but then again the danger was that much greater. It was one thing to try and drive a jackal off a carcass, but the group would steer clear of predators like the false saber-toothed cat or the giant hyena, or even the humans.

But the worst were the leopards. The female that had been dragged off by the leopard was the second loss the group had suffered in the past weeks. Earlier, the youngest of his brothers had been killed by the cat while they were looking for food on the same grassy slopes above the riverine bush. The attacks had been traumatic. In both cases the group had been taken by surprise. The leopard had followed them through the undergrowth, waiting for an individual to stray away from the others. Then it had struck in a flash, springing from behind the tall grass, seizing the unsuspecting victim by the neck. The next 60 seconds were a confused blur of dust and blood and motion, punctuated by the terminal blood-curdling scream of the terrified victim. By the time the group had realized what had happened, the victim was dead and being dragged off back into the grass. Normally, if confronted by a carnivore, the group would make a collective screeching noise, stamp the ground, wave their hands, and throw stones, a display that often forced the attacking beast to retreat. This was also an effective technique for driving the smaller predators off a kill. But this time it happened too fast. The leopard hauled its lifeless victim back to a cave a few hundred yards away, where it indulged in a macabre feast over the next three days.

The group now avoided the slopes, having discovered that the leopard occupied a cave on the hillside. Instead, their foraging efforts would follow the watercourse until it opened into a broader valley, where the trees thinned out and the grassland unfolded over the dolomitic outcrops. There were several fruiting trees in the vicinity, enough to keep the group happy for the next few hours. He signaled for them to stop. He and his brother then left the group and walked toward a nearby ridge to survey their food route. As always their progress was slow; every few paces they would halt to either ferret for grubs under stones or chew on leaves or seeded grass stalks, never missing an opportunity for food. He looked up, searching for the presence of vultures, which would indicate a kill they could try to steal. But today there was nothing. Farther down the valley, they noticed a number of grazing animals—giant buffalo and several antelope. This meant there were no predators around, a state of affairs confirmed by the number of baboons sunning themselves peacefully on the rocks at the edge of the valley where the foothills began.

His attention was drawn by his brother, who had come across a termite mound behind a rocky outcrop. They assembled around the sun-hardened dome to see whether it was still in use. There was one easy way to find out. He looked around for a stick or a piece of bone that he could use as a digging implement. It took him some time to find the appropriate instrument, a piece of long bone shaft broken into a sharp sticklike object by the jaws of a hyena. He used the bone to pierce the side of the mound. Chopping in quick hard strokes, it didn't take him long to reveal a mass of scurrying insects. Throwing the bone aside, he picked up a long blade of grass and licked it through his lip, coating it with a slick layer of saliva. Laying it into the mass of termites resulted in dozens sticking to its gooey surface. He quickly wiped it through his mouth, eating them all in a single motion, then repeated the process again and again.

Just a mile away, the bones of the two victims that the leopard had taken earlier lay in the cave long after the leopard had moved on. Initially other animals came and chewed on the rotting remnants of the leopard's meal. A porcupine dispersed most of the bones, nibbling for the nutrition of any leftover marrow and scattering the shards across the floor of the cave and toward the entrance. After several years, the only traces of the victims were the female skull and a fragment of the male jaw, which remained side by side, undisturbed in the tomb for the next 1.5 million years.

"Hey, what have we here? Romeo and Juliet?" I asked Andre Keyser as we stared at the extraordinary scene before us. Lying on its side was the complete skull and mandible of a female robust australopithecine, teeth gleaming bright white in the sun; upside down but immediately adjacent to this specimen was what was clearly a large male mandible. The skull was the most complete hominid skull I had ever seen, in or out of the ground.

But Andre is more of a classicist than I am. Later he and graduate student Colin Menter, Andre's assistant at the site, would name this remarkable pair Orpheus and Eurydice. I kidded both of them for being such dramatists, but I had to admire the appropriateness of their vivid imagination. Orpheus and Eurydice are characters from Greek myth. Orpheus, who had been given the gift of music and divine poetry, had fallen in love with Eurydice, but tragedy struck moments after their wedding ceremony when she was bitten by a snake and died. A grief-stricken Orpheus traveled into the underworld to try and find her and successfully bargained with the gods Hades and Persephone to allow her to return to the living world. There was a condition. He had to journey down a particular path without looking back over his shoulder. Eurydice would follow. Just as he was nearing the end of his path out of the underworld, he couldn't resist seeing if she was in fact behind him and turned, only to see her brief fleeting image vanish forever. Orpheus died soon afterward, attacked by women at a Bacchic orgy.

Thinking back now on the crushed but surprisingly complete female skull embedded in the rock with the male mandible lying adjacent, I realize just how fitting the titles of Orpheus and Eurydice are for these fossils, and I can understand how a find like this can bring out one's deeply lyrical instincts. The conditions set for Orpheus to lead Eurydice back to the world of the living would be difficult for any human to fulfill. Human curiosity is innate. Orpheus was not allowed to look back to where he had come from. My job is to do exactly that. The two beautiful specimens in front of me that day raised the gooseflesh on my arms as I realized I was one of the first humans to witness their extraction from the underworld.

A vivid imagination is both a curse and a blessing to the paleontologist. It is also a necessary tool of the trade. Even the most competent anatomist needs more than a scientific eye to visualize the flesh, muscle, and sinew wrapped around a fragment of fossilized bone to bring it to life. It is one thing to try and recreate the 1.5-million-year-old environment of the robust australopithecines as described above, but one has to be exceptionally careful not to start using the science to justify a preconception. In this case, too much imagination can be a bad thing, allowing muddled thinking to distort the relationship between objectivity and subjectivity and fuzz the science.

Dart discovered this to his detriment when he created his osteo-dento-keratic (ODK) flight of fancy. It may be academically forgivable to mistakenly believe that the blackened fossil bones of Makapansgat were damaged by a fire-wielding ape-man. But to claim that we are all psychologically indoctrinated prisoners of a two-million-year-old bone, tooth, and horn culture—and that humanity will remain that way for eternity—is beyond the bounds of acceptable scientific speculation. Much of Dart's reputation from Taung was undone by his insistence that his Makapansgat ape-man, *Astralopithecus prometheus* (later reclassified as *A. africanus*), was obsessed with hooves and horns because they were symbols of strength. "This strength," he wrote, "could not only be extracted from the animal but be assimilated to their victors, and be turned back upon the animals themselves for their undoing. The intellectual ability and manual dexterity displayed by the performance of these feats formed the background of their promethean culture."

Although the smashed bone fragments that Dart believed were the work of *prometheus* were actually caused mainly by hyenas, the great professor may have been inadvertently on to something. It appears that some of the ape-men species did in fact fashion rudimentary tools out of bone. This shouldn't surprise us. If modern-day chimps are capable of using makeshift tools to get food, why not the australopithecines, who were supposedly more advanced than apes? Whether they could make stone tools is another question entirely. By the time the robust australopithecines were at their evolutionary peak 1.5 million years ago, the art of stone toolmaking had been around for a million years. If the gracile australopithecines, who predated the robusts, were the earliest stone toolmakers, then surely the robusts would have been as well? The jury is still out, but from the evidence available it would appear that the robusts did not have the mental skills to make stone tools, although they may have used stone opportunistically in defense or in food gathering.

We've given the robusts, and the other australopithecines, a rough ride. In particular, the robusts are usually thought of as an unsuccessful species, a failed experiment in protohumanity because they traveled down a dead-end evolutionary road to extinction and vanished around one million years ago. But it is extremely presumptuous of us to regard them as failures. After all, they were on Earth for a lot longer than we have been. *Homo sapiens sapiens* have been in existence for between 100,000 and 200,000 years, during which time we may have risen to dominance over most other life-forms, but at the potential cost of our own survival. Our success is that there are over six billion human individuals on the planet, and although most live in poverty, their way of life is infinitely better than that of their predecessors of, say, 10,000 years ago. But in attaining control over our environment, we have depleted this planet's resources and developed the capacity to destroy ourselves several times over. What odds would you give that humanity as we know it will perish within the next 10,000, 50,000 or 100,000 years? The chances are fairly high, whether it be through disease, nuclear catastrophe, or genetic experimentation, to name just a few.

We should be therefore a little more circumspect when we talk of the robust australopithecines as being a bunch of evolutionary losers. Their species enjoyed a 1.6-million-year tenure on Earth before their ecological niche closed over, whether it was through competition with emerging *Homo* or an environment that shifted beyond their ability to adapt to it. That is nearly a million and a half years longer than our species has been around.

The robusts are nevertheless an evolutionary enigma. Possessing massive teeth and jaws, and with the males sporting an imposing crest running down the center of their skull, their appearance must have been startling. There are three known species, and they have one major feature in common—huge teeth behind the canines. Their molars are gigantic, with some individual teeth nearly the size of a U.S. nickel. They also all tend to have multiple roots in the premolars so the enormous teeth can be held in the jaw, whereas modern humans generally have a single root. Their jaws were massive in order to create the support structure for their large grinding platforms. In stark contrast to their huge back teeth, their front teeth are surprisingly small.

Why did they have this kind of dentition? The simple answer is that it was diet related. It would appear that the robusts ate a great deal of vegetable matter—roots, tubers, and seeds, foodstuff low in nutritional value and containing lots of grit. Their dental apparatus was modified for a nibble-and-chew lifestyle, which required enormous teeth to process this rough food and well-developed chewing muscles with large attachments along the side of the skull. Consequently, they had to have a massive skull to hold all this together, yet their cranial capacity would have been about one-third that of modern humans. Try and picture that, and you can see why imagination is an important part of reconstructing the past. The robusts literally had to chew their food all day to get enough nutritional value, and as a result their teeth began to wear very rapidly. Fossils show that they suffered from rapid tooth wear at an early age, and their life span would probably not have been much beyond 25 to 30 years, at which point they would have starved to death if not taken by a predator. In addition, the robusts were hugely sexually dimorphic, with the males often twice the size of the females. This tells us something about their social order and is the reason I used a gorilla-like model in the reconstruction. Males were probably extremely dominant, and a haremlike structure is the most plausible form of group structure I can imagine for such sexually dimorphic apes.

There has been extensive debate recently as to whether the three known species of robust ape-men should be classified as a separate genus because of their different anatomical characteristics. Many scientists are already doing so, categorizing them as *Paranthropus* (near man), which was the original taxonomy given to them by Robert Broom when he found the first of these species at Kromdraai in the 1940s. Superficially, this may seem justified. All the robusts share massive teeth and jaws and the traditional view of their evolution—defined by Johanson and White—is that their lineage arose from *Australopithecus africanus*. This was based on the fact that *africanus* had bigger molars than *afarensis*, and if one carries that logic through to the next step, it had to lead to the robusts. This certainly appears to be the case with the South African variety of the robusts, *Australopithecus robustus*. Ron Clarke's facial reconstructions of gracile australopithecines from Sterkfontein show signs that their molar development is on an evolutionary pathway to becoming robust. The problem is that the oldest East African robust species, *Australopithecus aethiopicus,* appears to have been contemporary to *africanus*. It also would appear that the second East African species, *Australopithecus boisei,* is an evolutionary development directly from *aethiopicus*. *Aethiopicus* and *africanus* are such different creatures that the proponents of *Paranthropus* are left in a quandary—both could not have given rise to a single genus. The problem disappears if we accept that the East African and South African robusts are not related, but problems like this don't often go away so easily.

This scenario that the South African and East African robusts have a different ancestry is feasible if one considers the variety of hominids stalking the Pliocene landscape. If the climatic changes that began three million years ago prompted drier conditions across the continent, then it is arguable that two different types of hominids responded in the same way to the changing environment, the encroaching savanna. One can see how hominids in different parts of Africa would have all developed bigger teeth as a survival mechanism in order to process new kinds of foodstuff. If the robusts are not all descendants of a single ancestral species, then they cannot all be referred to as a single genus, *Paranthropus*. In this case the title of *Australopithecus* remains intact because it was the stem species for all later hominids.

The robusts have always been considered vegetarians because of these big teeth, and diet has often been put forward as a reason for their extinction. This argument would hold that they weren't getting the kind of protein that the meat-eating *Homo* was, and therefore lost out in the race for bigger brain size. And because their brains didn't cross the proverbial cerebral Rubicon of 600 cubic centimeters, they were therefore incapable of manufacturing tools. Whatever the cause for the robusts's extinction, these two assumptions have been challenged by recent research. First, it appears that the robusts were opportunistic eaters, and therefore did have some meat in their diet; second, they probably were capable of making primitive tools, although not in stone.

The remains of bone tools have been found at both Swartkrans and Drimolen. Bob Brain was the first to recognize these tools back in 1979. During the next seven years, he found almost 70 bone tools at Swartkrans, but because of the prevalence of both *Homo* and robust fossils in the same sediments, the question of who actually made them remained inconclusive. At Drimolen, also a great sample of robusts, Andre Keyser has found bone tools. Because relatively little is known about the postcranial anatomy of the robusts, however, he has also found it very difficult to distinguish *Homo* remains from robust remains. But an advance has been made in postcranial analysis by Randall Sussman of Stonybrook, who studied robust hand bones from Swartkrans. He believes there is at least the possibility that the robusts were physically capable of making complex tools, although whether they actually did is another matter entirely.

Another study of note was conducted by my colleague Peter Unger of the University of Arkansas. Using studies of micro-wear patterns on the dentition of robusts, he has concluded that robusts must have had some meat in their diet, but they were largely vegetarians. So were they or were they not tool users and makers? Were the so-called bone tools the product of robust australopithecine culture, or were they just mimics created, like Dart's ODK, by the actions of hyenas?

These were problems that I put to Lucinda Backwell, a graduate student working with me. Lucy, as she's also known, which of course is a very fashionable name in paleontological circles, attacked the bone-tool issue with vigor. I had my doubts as to whether the Swartkrans bone tools were in fact tools at all, as the markings on them are unlike any others that exist in the archaeological record. Bob Brain and Pat Shipman from the Pennsylvania State University had studied these bones under electron microscopes and found them to have wears and scratch marks, but I didn't believe these were necessarily caused by hominids. The marks looked very similar to hyena-damage marks studied by Paula Villa in ancient hyena lairs in France. Backwell was aware of my skepticism and knew that I was hoping that she would disprove Bob Brain's claims, much as he had disproved Dart's decades before.

Backwell began her research by selecting data on every known accumulating agent that has been identified in Africa. Accumulation agents need not only be live animals like leopards, hyenas, dogs, hominids, and so on; they can also be inanimate forces of nature like water or river gravel. Backwell went through over 40 different accumulations in South Africa alone, each representing a different type of accumulating agent. In the course of this painstaking research, she literally looked at tens of thousands of bones collected by animals or modified by geological processes.

What she found was a series of what we called pseudo-tool bones, which closely mimicked the so-called bone tools from Swartkrans, but were in fact the result of natural damage rather than shaped by hominids. If this was the case, Brain was wrong, and the tools he discovered were actually damaged bits of bone that were not fashioned by hominid hands.

As Backwell got down to the microscopic details of the study, however, it appeared that Brain may have been right after all. Using scanning electron and confocal microscopes, she found that the scratch marks that appeared on the Swartkrans bones did not appear on any of the other pseudo-tools from the tens of thousands of bones she had examined. They were unique. Maybe they were bone tools after all.

Backwell cast her net wider and traveled to Europe to examine bone-tool collections in Britain and France, and in the course of her travels enlisted the help of Francesco D'Errico. D'Errico is a Parisian who has become a world expert in the microscopic modification of bone surfaces. He accepted our invitation to come to South Africa to study the problem firsthand. During his ten days in the country, he and Backwell visited Swartkrans and Drimolen and studied their bone tools intensively. There was no doubt that the scratch patterns were unlike anything else D'Errico had come across in his vast experience. But he was stuck for an answer as to what could have caused the markings. Certainly nothing in the more modern bone-tool archaeological record even remotely resembled the radiating scratches on the bone tips.

The forces of chance came into play once again during a field visit when Backwell noticed a number of termite mounds near Swartkrans. Something flashed in her mind about chimpanzees using twigs to stick into termite mounds to get at what seemed to be a nutritionally rich delicacy. This gave her the idea to try out something similar with bones. She and Francesco got some bone flakes together and chopped into the termite mounds themselves. They then put their newly fashioned bone tools under a microscope, and lo and behold, found exactly the same radiating patterning of scratches that they'd been puzzling over for months.

The implications of this are significant. Backwell and D'Errico may have stumbled upon an extraordinary insight into the lifestyle of the robusts. For the first time, there is a linkage between the dietary evidence and the wear on robust teeth, the hand morphology that Randall Sussman showed was capable of tool use, and most importantly a level of tool use that reflected a behavioral pattern slightly more complicated than that of chimpanzees. Instead of using sticks as termite wands, they ground down pieces of bone on hard rock to make them suitable for getting at the little insects. The fact that termite mounds are found predominantly in open woodland and grassland areas confirms that these hominids had moved away from an arboreal lifestyle. It also explains the excessive wear on their teeth—separating the sand from the mounds and the termites could not have been an easy task, and in the process of eating termites, the robusts must have chewed a lot of grit. Perhaps this also can explain the kind of specialized niche that the robusts occupied while coexisting with a more sophisticated tool user like *Homo*.

Why then did the robusts ultimately perish? They didn't run out of termites, that's for sure. The conventional wisdom has been that their reliance on a vegetarian diet led them to extinction, that they could not compete with the bigger brained, meat-eating *Homo*. Early *Homo* would have therefore been smarter at getting food, reproducing themselves, and occupying all the better habitats, such as sheltered areas close to water. Although diet is a key factor in evolution, it would be simplistic to say vegetarianism alone was the downfall of the robusts.

The answer may have been that the robusts became extinct because they were *too* successful as a species. As contradictory as this may sound, it takes into account the changing nature of the African environment. Consider that both the robust and

Homo adaptations of early hominids were in response to the aridification pulse of the mid-Pliocene three million years ago. The robust response was one of adapting to eating the low nutritional plants of the broken woodland and savanna. Yes, they were opportunistic in eating anything they could, but their dietary mainstay was plant material. *Homo* on the other hand was more of a generalist, depending on scavenging across a variety of habitats for food. It follows, then, that if the robusts were so dependent on a specific environmental niche, they faced decline if that environment changed, or if it became very competitive.

The robusts may well have been squeezed out of their environmental niche by ancestral baboons on the one hand and early *Homo* on the other. Baboons today forage for almost exactly the same food as the robusts probably did, while the more superior *Homo erectus* would have been a better scavenger and hunter. Add to this that the robusts were probably fairly easy pickings for carnivores: Bob Brain's work at Swartkrans shows that a large number of the robusts ended up as leopard food. In the course of time, they appeared to get pushed to the edge of the hominid food chain and simply could not keep up a viable population. It is a question of simple arithmetic that if your mortality rate continually exceeds your birthrate, there is no doubt that you are heading for extinction—it is just a matter of the time it takes to get there.

This raises a point about extinction. In the popular mind it is often a sudden apocalyptic event that precipitates death and destruction in one dramatic flash. This may well have been the case with the dinosaurs. There is significant evidence that a mass extinction of most land animals occurred 65 million years ago, probably because of a gigantic meteorite strike in the Gulf of Mexico. The impact is believed to have led to dustlike clouds encircling the globe, blocking out the sun's rays and halting photosynthesis in plants, which in turn affected the food chain. Those creatures that had the biggest food requirements did not survive. But in most cases extinction is more subtle. As habitats change and the food chain shifts, species have to adapt or die.

The day after first seeing Orpheus and Eurydice, I assisted Keyser and Ron Clarke in removing the fossils from the rocks, carefully using our brushes and dental picks to clear the layers of dirt. One was immediately struck by the difference in size—the female jaw would have fitted entirely within the male mandible; the teeth of the female were about half the size of the male. Clarke was very excited. He took the specimens back to his lab to undertake a reconstruction. This was in late 1994. Over four and a half years later, we were still waiting for Ron to finish.

Andre Keyser holds a cast of a male mandible of a robust australopithecine (left) and a female boisei *skull (right) directly above the mandible of Orpheus and the skull of Eurydice, both still embedded in the rock prior to their excavation.*

I had known Keyser was on to something at Drimolen since August 1992, when he called me from Pretoria to say that he had recently begun an examination of a cave deposit near Sterkfontein. On only his third visit to the site, he had discovered a hominid tooth in the wall of breccia projecting from the ground. The tooth turned out to be from a robust australopithecine and was followed rapidly by several other specimens.

In just over a year and a half, Drimolin would prove to be one of the richest of the newly discovered sites, revealing literally dozens of fossils of robust ape-men and even a few remains that would be attributed to the genus *Homo*. Among the more unusual fossils that have come from Drimolen have been the remains of a year-old *Homo* infant—to date the first in the world —as well as several juvenile *robustus*.

There is still a lot we have to learn about the robusts, which in South Africa have been found at five sites, all in the northwestern vicinity of Johannesburg: Kromdraai, Swartkrans, Gondolin, Coopers, and Drimolin—the latter two being the newest robust sites. Kromdraai is the original site where Broom found the first robust. Swartkrans is the most famous, having yielded over 100 individuals. Drimolen probably has the most potential, having provided us with over 80 hominids, including the most complete australopithecine skeleton thus found. It also has lots of babies. Coopers is a relatively new site that I have been digging with another graduate student of mine, Christine Steininger, who found the face of a robust overlooked during excavations by the Transvaal Museum in the1950s.

But from about 1.7 million years to the time of the extinction of the robusts at about 1 million years ago, these large, flat-faced ape-men aren't the only players on the African scene. The other is of course our ancestor: *Homo erectus*.

One of the best examples of *erectus* is the Turkana Boy. I will never forget my first view of that skeleton, shown to me by Meave Leakey during my visit to Kenya. As she pulled open the drawers in the Kenya National Museum hominid strong room, I had a sensation akin to the breath being sucked out of my body. Having worked with australopithecines so long and having been so engrossed in their apish morphology, it was emotionally striking to see such a humanlike fossil. From head to toe, it was clear that by 1.7 million years ago, regardless of what had gone on before, hominid evolution had settled on a basic body plan that was remarkably similar to the one we have today.

After *erectus* appears in Africa around 1.7 million years ago, we rapidly see a depletion of variation. All of the other species of *Homo* apparently disappear; only the robusts survive the emergence of *Homo erectus*. You have to imagine an animal that has the stature of a modern human but a brain only about 75 percent as large. This was a maker of complex stone tools, handaxes, choppers, and spheroids, all designed and used for specific tasks. But members of this species were not humans. They lacked the ability to develop a more complex tool kit. They lacked, apparently, the ability to think symbolically like a modern human. They did not bury their dead or perform other complex rituals. They were effectively human in physique but not in mind. Alan Walker has described, in possibly the most effective and visual way, what these protohumans must have been like: They were the velociraptors of the Plio-Pleistocene.

Some time between 1.8 and 1.3 million years ago, *Homo erectus* succeeds in spreading out of Africa and occupying much of the Old World. Around 1.1 million years ago, we see the extinction of the robust lineage's in both eastern and southern Africa corresponding—perhaps not coincidentally—with the evidence of the first controlled use of fire at Swartkrans in South Africa. And by one million years ago, populations of *Homo erectus* are in the Far East, Asia, and Europe, and of course widespread in Africa. But for the next million years, while we have a good fossil record of human evolution in the Near East and Europe, and a fossil record of *Homo erectus* in the Far East and Indo-Pacific, the fossil record in Africa is almost nonexistent. While stone tools litter the landscape of Africa, indicating plentiful populations of hominids, we have only a tiny handful of fossils from across the continent. The record is so poor that we have begun to call this period in Africa the Million Year Gap.

CHAPTER 12

FOOTSTEPS OF EVE

～

The tearoom of the archaeology department at the University of Cape Town has a spectacular view. Perched on the slopes of Table Mountain, the elegant, ivy-covered stone buildings have a vista that stretches northward, over the housing estates of the Cape Flats to the dramatically sculptured mountains that fringe the region's famous winelands. To the right, 20 miles away, is False Bay, which arches in a broad sweep from the Peninsula mountains, framing the villages of Kalk Bay, Fishoek, and Simonstown, through to Cape Hangklip, a distinctive dome-shaped promontory barely visible over the ocean haze. To the left, masked by the other campus buildings and the slopes of Devils Peak, is the Indian Ocean and Table Bay, a harbor that cuts into the lip of the bustling city bowl. On a clear day one can just make out Robben Island, the sliver of a landmass 11 miles beyond the harbor that for almost two decades was the prison home of Nelson Mandela.

Cape Town's relationship to these two oceans is its raison d'être. Positioned at the bottom tip of Africa, the early Cape was a refueling point for the European voyagers en route to the Far East. Strategically positioned, it was successively used by the Portuguese, Dutch, British, and French in an attempt to control the sea lanes around the continent. This patchwork colonial heritage—ethnically enriched by the influx of slaves from the Far East, the brown hues of the indigenous Khoi and San peoples, European sailors, and the darker tones of Africans who've migrated from farther north—gives Cape Town a spicy, cosmopolitan feel. Yet flat-topped Table Mountain in the distance dominates the skyline, dwarfing a city characterized by a vibrant and troubled history.

I've often caught myself staring out of the tearoom window like some armchair philosopher, wondering what it is about this part of the world that has played such a role in shaping humanity. South Africa is the product of a tortured past. In contemporary terms it is a curious country, essentially a conservative nation caught in a tide of radical social change. It is the human equivalent of the San Andreas Fault, defined by and yet defying racial fractures that run through the political landscape like geological schisms. As anarchy and opportunity jostle for center stage, hope and despair are worn and shed on a daily basis like clothes. At any moment one feels South Africa could slide into the social chasm that afflicts so many other African countries to the north. Yet at the same time, its convoluted commitment to racial reconciliation could just prove to be a model for the 21st century. There's a surprising depth to the subcontinent, a beguiling overlay of the new and the old, in which adolescence and ancient wisdom have somehow found a comfortable fit. It's people are surprisingly traditionalist, and yet in this milieu some of the most radical ideas have taken root.

When we consider the notion of history, we get far too caught up in the concept that it has been molded by human action. We believe that the world is shaped by individuals or groups of people acting either in concert or against each other, and that the consequences of these actions constitute the reality of the present. But when one reflects on a deeper history of humanity, the factors influencing change are almost all extraneous. Essentially, we are what we are because of the environment we live in. And when that environment changes, so do we. This has been the driving force in evolution, and our responses to the shift in climate and the available diet determines what we are today.

The Earth is a dynamic body. I'm reminded of this every time I soak in the marvelous view from the Cape Town archaeology department and imagine what it was like when the lowlands in front of me were covered with water. Barely 10,000 years ago, Table Mountain and the southern peninsula were cut off from the rest of Africa by the sea, which spilled across the Cape Flats. The sea levels, which we take for granted today, have been around for only about 3,000 years. The ebb and flow of life is reflected in the way our continents move and our coastlines shift. During the Miocene, between 13 million and 5 million years ago, sea levels were some 400 feet above their present levels. Should humanity survive the next 10,000 years, the odds are that the world's famous coastal cities—New York, Hong Kong, Sydney, London, and even Cape Town—will all be underwater archaeological relics.

In 1993, I spent a semester at the University of Cape Town as a visiting lecturer in paleoanthropology. The assignment offered me time to finish writing my Ph.D. thesis in a relatively relaxed environment, away from the jostle of Johannesburg, a rowdy frontier mining town by comparison. I'd also become intrigued by the Cape's remarkable fossil record, which, through the apartheid years, had remained out of the glare of international scientific attention. From the archaeology department, I had often gazed over my morning cup of coffee toward the naval base at Simonstown. This was where a plucky wooden frigate called the HMS *Beagle* docked one fine May morning in 1838 after six months at sea. One of the crew members who would gladly have stepped ashore was the ship's young naturalist, Charles Darwin. He had been traveling the world, gathering the information that he would eventually consolidate into his overarching theory of evolution. Darwin had a strong hunch that humankind had originated in Africa, as is evident in this frequently quoted passage from his *The Descent of Man* (1871):

> In each great region of the world the living mammals are closely related to the extinct species of the same regions. It is, therefore, probable that Africa was formerly inhabited by extinct apes closely allied to the gorilla and the chimpanzee; and as these two species are now man's nearest allies, it is somewhat more probable that our early progenitors lived on the African continent than elsewhere.

At the time, however, Darwin was taken more by sociopolitical issues than issues of animal ancestry. In particular, he was highly critical of slavery and the racism he perceived among the Christian missionaries, and the general ill-treatment of indigenous populations. He admitted, however, that he found it hard to see how primitive islanders, such as the "barbarous savages" he came across at Tierra del Fuego, could have shared the same lineage with more "civilized" Europeans.

The *Beagle* docked in the Cape on its journey from Australia to Brazil via stopovers at the Indian Ocean islands of Cocos and Mauritius. Like most other visitors who alight in the shadow of Table Mountain, Darwin was impressed by the majestic landscape of the peninsula and Cape Town itself, where he stayed for a month while the ship was reprovisioned. Surprisingly for a naturalist, but not so to anyone who has savored the pleasures of Cape Town, Darwin made little attempt to explore the landscape beyond the colonial comforts of the city. In fact, he appeared to be intimidated by the countryside, abandoning a trip into the interior after confronting the barren stretches of the Karoo, the arid semi-desert that separates South Africa's western coastal plains from the rest of the country. He wrote later that he found this to be a bleak and depressing landscape with little to recommend it.

Had Darwin the time and the inclination, he would have found that the Karoo is a naturalist's heaven. It is an enormous basin, exposed by the retreat of the ice age that had covered most of southern Africa in an ice cap between 250 and 190 million years ago. It is difficult today—as it would have been for Darwin —to imagine that this rugged and inhospitable region of stone and sand was once a temperate, lush swampland, irrigated by periodic heavy rains before being turned into a wasteland by volcanic activity around 190 million years ago. During this time, the world's first mammal-like reptiles, ancestors of the mammals, and archosaurs, ancestors to the dinosaur, roamed these marshes, leaving a myriad of fossils to which scientists like Robert Broom and James Kitching hitched their fame. More important, the Karoo and the adjoining thirstlands of the Kalahari were barriers that would have isolated relatively small populations of archaic *Homo sapiens* and may have allowed the physical and behavioral development characterizing the cultural foundation of our species today.

One Saturday morning the head of the archaeology department, John Parkington, invited me to go and look at a site that he was interested in digging on the western Cape seaboard, a few hours north of Cape Town. Parkington, a relocated Scotsman, is a specialist in the spatial patterning of archeological sites of the Middle and Late Stone Ages. As we drove up the West Coast, Parkington explained the valuable insights that this region offers, not only into early human prehistory but also into the emergence of modern forms of African mammals. The fossil beds in the area around Langebaan also give us a particularly rich insight into the Early Pliocene.

The first fossils discovered in South Africa were found 100 miles north of Langebaan, near a small fishing village called Port Nolloth. They were described in 1779 by W. Paterson, who was traveling on horseback with his companion Mr. R. J. Gordon when "several petrifications of shells, some of which were about an hundred and fifty feet above the surface of the sea" were found. Little did Paterson know, the *Ostrea prismatica* fossils are now recognized as markers of raised-beach deposits containing diamonds. Had they been aware of the treasures that lay beneath the surface on that inhospitable coastline, he and Gordon could have been wealthy beyond their wildest dreams.

That area is now known as the Skeleton Coast and contains some of the world's richest diamond deposits. The fact that the shells were so high above the existing sea level is indicative of how dramatically the South African coastline has changed over the past millions of years as the sea rose and dropped in the process of the Earth's own evolution. These same shells have now acted as markers for our research along this same coastline.

The West Coast is also remarkable for its later fossil heritage. This first came to light in the late 1940s, when fossil bones from a farm called Elandsfontein in the Hopefield district near Saldanha Bay were brought to the attention of academics at UCT. It was only in the 1950s, however, that Ronald Singer, a senior assistant in the department of anatomy at the university managed to get to the site for a reconnaissance. Confronted by 50-foot sand dunes that were continually scoured by the Cape's well-known fierce southeasterly winds, Singer found a profusion of stone tools and bones. Realizing that Elandsfontein was an important site, he recommended that a more detailed scientific study of the area be undertaken.

Three years later, Keith Jolly, who'd been appointed the field officer for the site, came across fragments of the famous Saldanha skull. His discovery was assisted by the relentless southeaster, which exposed a number of bone fragments that he saw while walking through the 2,000-acre area described by archaeologist H. C. Wodehouse as an "ever-moving, unmapped, fossil-bearing field." Jolly recovered 25 bone fragments, which eventually fitted together into a skullcap displaying heavy brow ridges that were primitive yet more derived than *Homo erectus*. Subsequently, the Saldanha skull has been dated to about 500,000 years before present, placing it in the middle of the million-year gap, and it seems to represent an early form of "archaic" *Homo sapiens*, the first of our kind.

Elandsfontein eventually yielded close to 100,000 fossil bones and stone artifacts, enabling us to build up a picture of the environment of the Middle Pleistocene (between 800,000 and 400,000 years ago). Among the fossils found were the remains of *Hipparion*, the three-toed horse ancestral to zebras and horses of the genus *Equus*; the giant baboon (*Therapithecus oswaldi*), which is twice the size of the baboons found today at Cape Point; the short-necked giraffe *Sivatherium*, with antlers like a moose; and a saber-toothed cat. The region was at the time certainly more habitable than it is today, and probably consisted of mixed woodland thicket that allowed far more animal diversity than currently exists. An indication of this is that there were up to six members of the wildebeest-hartebeest family, which are dedicated grazers; there are only three today, while browsers like the kudu and mixed feeders like the eland were also present. The bone fossils of larger herbivores, such as the extinct giant buffalo, hippos, black, and white rhinos, elephant, pigs, giraffids, and carnivores, have also been found during numerous field seasons since Jolly's initial discovery.

More recently research conducted by the South African National Museum has focused on Langebaanweg, about five miles due east from Saldanha Bay, where the fluctuations in sea levels that accompanied the shift from the Miocene into the early Pliocene left a mass fossil grave of a variety of animals. The fact that tropical animals needed to stick closer to water made them all the more vulnerable to predators or drowning. An astonishing 200 species of vertebrate and invertebrate fossil animals have been recovered from an ancient riverbed, including 80 mammals and 60 birds, victims of rising sea levels caused by the melting of the Antarctic ice sheets. In some cases, flash floods caught groups of animals by surprise, burying them upright in river sediments.

In one instance, the remains of about 500 large giraffe-like creatures (*sivatheres*), some weighing as much as a ton, have been found in a single location, obviously taken unawares by the raging waters. There are other oddities: In the space of one square yard, there are the bones of 800 individual golden moles.

These remains tell a dramatic story of the major ecological transition that took place around five million years ago as the continent began drying out and the great tropical forests retreated in the face of an advancing savanna. Although primate fossils are almost non-existent at Langebaanweg, they would have been affected by the same aridification process in which forest-dependent creatures become increasingly isolated in patches of dense vegetation along the riverbanks as water became scarcer and scarcer.

It is, therefore, no wonder that Parkington developed a keen interest in the West Coast. The site that was of particular interest to him and that he wanted me to take a look at is called Hoedjiespunt (Little Hat Point), which sits on a spit of land jutting out into the Atlantic Ocean. To get there one has to drive through the town of Saldanha, a rather ramshackle village, quaint inasmuch as the local semi-industrial fishing industry will allow. Despite the peri-urban economic activity that takes place at Saldanha, it still manages to convey a sense of remote bleakness, heightened all the more by a discomfort caused by the continuous glare of sunlight off the white sand dunes that dominate the coastline.

On the drive from Cape Town, John explained that he believed he had found a human occupation site from the Middle Stone Age, where people had sat on a sand dune around a hearth and eaten mussels and limpets (a type of shellfish) as well as seals, turtles, and small antelopes. Judging by the tools found, Middle Stone Age implements, he believed that this event took place between 200,000 and 45,000 years ago. The site was in good condition, probably because it had been cemented by calcium carbonates leaching from the dune after a rapid burial in sand. But while preserving this wonderful event in time, the calcification made excavation exceedingly difficult. John wanted my opinion as to whether he could use techniques similar to those we used to extract tools, bones, and shells from the hard dolomitic breccias of the Witwatersrand.

John also idly mentioned that the site was near a hyena lair where Richard Kline, the famous faunal analyst who had worked on nearly every Cape fossil collection in existence, and Graham Avery, director of archeology at the South African Museum in Cape Town, had made a small collection of bones back in the 1970s. These had been dated, based on faunal comparisons, to about 74,000 years before present. If he could tie the two sites together geologically, this might indicate that the archeological site was of a similar age.

These thoughts were going through my head when the old Land Rover engine sneezed and spluttered into silence as John pulled up to the seaward side of a dune, which was exposed to the pounding Atlantic. He and I, his assistant Cedric, and a number of graduate students including Deano Synder, who would eventually work with me exploring Botswana, stumbled out of the vehicle, relieved to stretch our legs after the two-hour drive from Cape Town. Thirty feet below us, the waves crashed onto the shoreline. Shielding our eyes against the harsh glare, we began walking up the steep slope to the fossilized hyena's lair that John had located at the top of the bleached white dunes. I marveled at the thought that people had actually sat on this dune in the Middle Stone Age, roasting oysters and other seafood delicacies in much the same style as some of the picnickers we noticed on our drive.

About halfway up the slope, John started pointing out bits of bone eroding from the sand and presumably originating from the ancient hyena lair. A particularly nice bone caught our attention and we both kneeled down to look more closely at this fragment. As I bent down, I happened to look where my knee had landed and noted a small glint of bright white enamel about an eighth of an inch long. It was clearly and unmistakably the central part of a human tooth. "This is a human tooth," I said to the group. Everyone stopped in their tracks and looked at me with disbelief. Hominid fossils from the Middle Stone Age are exceptionally rare; the tiny fossil fragment I was holding in my hand needed some imagination to morph it into anything that resembled a tooth. "Well, it's not a whole tooth but it's certainly part of one. See how thick is the enamel?" I insisted to my immediately skeptical audience. John Parkington looked at me like I was having him on.

Cedric took a more proactive stance: "I think it might be a pig," he ventured, looking carefully at the tiny fragment. The thick enamel of pig teeth has fooled more than one paleoanthropologist into thinking he's found a hominid.

"I'm sure it's not," I replied. "It's primate and it's got to be human." I held the fragment close to my face. See the central fovea, the depression down the middle?" I pointed to the small valley. "That's hominid for sure."

"If you say so," said John. "We had better see if there's more." We began a careful search of the immediate area. Within a few inches of where the first piece was found, we recovered another shard of thick enamel. Our original mission, to look at the human occupation at the top of the dune, was forgotten. We spent the rest of the morning scouring the surface of the hyena lair. Eventually I had 13 fragments of what appeared to be the same tooth—a thimbleful of enamel that I was certain was human.

"I still think it's a pig," said Cedric.

The next day, back at the University of Cape Town lab, I sat down in front of a microscope to put together what seemed like the world's tiniest jigsaw puzzle. After several hours of painstakingly fitting together the fragments, using tweezers and glue under the magnification of a binocular microscope, I was eventually rewarded with success. The puzzle was completed to form a clearly identifiable human upper molar. It showed barely any signs of wear, indicating that the individual it had come from was young, perhaps only a teenager. I called John and Cedric and showed them the completed tooth, now clearly that of a human—the Hoedjiespunt Child, as I dubbed him.

Cedric smiled, "I knew it wasn't a pig."

We were puzzled by the tooth because it appeared to be too primitive for the 75,000-year-old estimation that John had given the site of the hyena lair, and because it resembled more the morphology of specimens attributed to much older hominids such as a *Homo rhodesiensis* specimen from Zambia, thought to be about 300,000 to 500,000 years old. Regardless, I was pretty sure even then that is wasn't a "modern" human tooth.

Our subsequent work at Hoedjiespunt over the next few years helped clarify matters further. During that time, we recovered more human fossils from the hyena lair, some teeth, even a few fragments of skull, and a very robust tibia or shin bone, all of which we believe belonged to the same individual who'd given us the tooth. Our hunch that the Hoedjiespunt Child was not a modern human was confirmed by these finds. Its skull is somewhat thickened and the inner lining of bone in the tibia, the cortical bone, takes up a greater proportion of the interior of the shin bone than one would expect in a living human, particularly a child. We suspect that the Hoedjiespunt Child is a member of a group of hominids that are often labeled archaic *Homo sapiens*. This is really just a convenient term for the muddle of fossils from Africa and Europe between about one million and 200 thousand years ago.

Working with a number of geochronologists, most particularly John Vogel and Stephan Woodborne of the CSIR in South Africa, we've been able to establish that the fossil-bearing sediments of the hyena lair were significantly older than we believed originally, dating to about 300,000 years before present. Interestingly, modern geneticists believe that modern humans evolved from this archaic form of our species around 200,000 years ago. This obviously begs the question as to what happened in that intervening 100,000-year period that gave rise to modern humans.

So-called archaic *Homo sapiens* seem to combine many of the robust features of *Homo erectus*: prominent brow ridges and a large face, with characteristics of our own species *Homo sapiens*, which has a more domed brain case and jutting chin. Although the evidence is not conclusive, it would appear that *Homo erectus*, or *Homo ergaster*, began taking on slightly more modern characteristics around 600,000 years ago and developing into the archaic version of our own species. One of the main problems with the African archaics, however, is the frustrating lack of specimens. In total, there are probably less than three or four dozen representative fossils of this group from all of Africa, and each varies slightly in its morphology from the others.

One of the best representatives of this group is the Broken Hill skull found in 1921 by workers at the Broken Hill zinc mine in northern Rhodesia (now Zambia). The miners were digging out a cave when they discovered parts of at least three skeletons. Bones from the arm, leg, and hip as well as a skull were found, but unfortunately, like many early discoveries, the precise context of the finds was never clearly defined. Many other specimens from what should have been a valuable site were lost through filching by souvenir hunters and the rather destructive nature of the zinc mining process. Nevertheless, the find resulted in the recovery of the most complete skull of an archaic *Homo sapiens* ever found. Enthusiastic researchers dubbed this creature Rhodesian Man.

Rhodesian Man's brain was surprisingly large. At about 1,300 cubic centimeters, it was well within the range of modern humans. From recent reconstructions of the height of Broken Hill Man using the tibia, it can be estimated that an adult would be close to six feet tall. Using the size of the muscle markings on the bones, we can also ascertain that the Broken Hill individuals would have been extremely well muscled. Interestingly, the teeth of Broken Hill Man are in appalling condition. Defects and cavities abound. It has been suggested that the individual died of infection spread from these bad teeth. More recent research has even suggested that the Broken Hill Man died of mercury poisoning (a potential cause of dental damage) from consuming very high quantities of the mercury that is prevalent in the soils of the area.

One of the better dated archaics discovered comes from Ndutu in Tanzania, where a skull similar to the Broken Hill specimen has been dated to between 300,000 and 400,000 years before present. This skull also had very heavy brow ridges but a higher forehead and more rounded cranium, a feature usually associated with modern human crania. In Ethiopia, at a site known as Bodo, an even more robust cranium than the Broken Hill skull has been found. This find is also dated to around 300,000 to 400,000 years old. Of the three major finds from South Africa, two—the Hoedjiespunt Child, and a skullcap from Saldanah Bay on the Western Cape—were dated to about 350,000 years ago, and a third—a skull and face from Florisbad in the Free State—recently was redated to about 250,000 to 350,000 years old (it was originally thought to be only about 150,000 years old). An unusual and interesting fossil attributed to this group is the Berg Aukus femur from Namibia, an enormous femur that may have belonged to an individual over 6.5 feet tall. Other than these finds, there are only a few other bits and pieces from around Africa. Yet despite this near absence of evidence, this must be one of the most crucial periods in human evolution. We have moved beyond the morphology of *Homo erectus* and sit on the verge of the emergence of anatomically modern humans. It is in these archaics' morphology and behavior that dwells the essence of Eve.

Despite this near absence of fossils, the archaeology, or the cultural remains from this period, is another matter entirely. The modern African surface is literally covered with the stone tools left behind by these mysterious near-humans who had made and used the same kind of stone tools as their predecessors, *Homo erectus*, the so-called Acheulean industry.

Acheulean sites were generally in valley bottoms or wetlands; this species was a terrain specialist, living in the riverine forests that traversed the plains. Acheulean tools are associated with *Homo ergaster, erectus,* and archaic *sapiens,* thus representing a tool type that spanned almost 1.5 million years of human evolution and three species of hominid. We classify the culture of the archaics as broadly within the Early Stone Age. The favorite implement was the hand ax, a technology that had existed virtually unchanged for the previous million years. Hand axes were in widespread use from Africa to southern Europe and as far east as India, indicating a widespread sharing of a basic knowledge. We believe that these large bifacial axes were handheld and were probably used as a butchering tool. Their sheer size indicates that the archaics were more muscular than we are today; the fact that the technology remained constant for such a long time indicates that the archaics were niche players, living successfully within specific terrains. From the position of most Acheulean sites, these were generally close to water, either in riverine forests or wetlands, where there would have been plenty of animals to either hunt or scavenge.

The richness, or at the very least the density of their stone-tool technology, is particularly evident in the drylands of the Northern Cape, south of the towns of Uppington and Kimberley, which would have been significantly more temperate environments during the Early and Middle Stone Ages. Just south of Kimberley, Peter Beaumont of the Kimberley Museum has found what must be one of the world's largest accumulations of stone tools at a single site. In a space less than that of two football fields, he estimates that there are over 10 million hand axes and other stone tools. Further west, near Uppington, there are literally tens of thousands of hand axes, choppers, cores, and other elements of this culture lying scattered across almost every erosional surface you chance upon. While this sort of density of tools is not common, the number of artifacts that one can find from this time period illustrates the success of these archaic *Homo sapiens*. Surprisingly, there are relatively few sites anywhere in Africa with good, undisturbed archeological sequences that we attribute to archaic *sapiens,* so our understanding of their behavior is limited.

Nevertheless, we can see from these few glimpses into the past that, like *Homo erectus,* the archaic *Homo sapiens* had a limited scope for complex thought processes. Their tool types were limited to maybe just over a dozen different kinds of stone implements. But they could certainly effectively control and possibly make fire. These limitations in tool diversity would indicate that communication between the archaics was not what we would call modern human language. It is reasonable to assume that the increasing sophistication of the Late Stone Age tool kit around 250,000 years ago, in which there is a greater variety of implements and a move toward more specific-use type tools, may be a reflection of the development of modern human language. The Middle Stone Age is the first time we see refined and hafted spear points being manufactured; there are scrapers, choppers, and small blades. And recent evidence from caves and open sites in coastal southern Africa suggests that by around 100,000 years ago, this culture gained the rudimentary aspects of modern human behavior. At sites like Blombos cave on the southern Cape and at Hoedjiespunt on the western Cape, the use of bone harpoons and ocher had become common. Bone harpoons indicate a sophisticated level of tool manufacture and food acquisition that is clearly modern. Ocher seems to indicate that paint was being used either for decorating objects or as makeup. The most startling discovery, and possibly the most controversial, is the excavation of a child's burial from Border Cave in KwaZulu-Natal, perched on the frontier of the Kingdom of Swaziland. The dates, according to a variety of techniques, indicate that this child was buried some 60,000 to 80,000 years ago, its small body covered in ocher before internment.

These seemingly modern behaviors at such an early stage are startling to many. Prior to these finds it was thought that this level of modern human culture would not appear until around 50,000 years ago; certainly the earliest evidence in Europe is not older than this. But something different was happening in the coastal regions of southern Africa. We just don't know what that was.

John Parkington and I have debated these points at length during my annual visits to the Cape, which I've taken since teaching there for a semester in 1993. We would make a regular pilgrimage up the West Coast to Hoedjiespunt in his battered Land Rover to continue digging the hyena site. During these excursions, we would rent a house in Saldanha Bay, where we'd talk late into the night about the exciting body of evidence in the search for human origins that was emerging from South Africa.

On one summer afternoon in 1996, after a particularly hot day at Hoedjiespunt, we returned to the rented suburban bungalow, tired and hungry. We opened cold beers and collapsed on the porch.

"There are some human footprints in the sand dunes near here that may be about 100,000 old," John Parkington said with a feigned casualness, motioning lazily toward the Atlantic coastline that stretched southward in a late afternoon haze. I almost choked on my beer. John grinned wickedly, obviously having carefully planned his statement for maximum effect. It worked. I wiped the dribble of beer off my chin and looked at him suspiciously.

"You're kidding."

"Nope," he said in his soft Scottish accent. "I've seen them."

John's bombshell distracted me from the enticing smell of the snoek that Cedric had just placed on the barbecue. I was in a state of disbelief. Fossils are rare; fossilized footprints are over a thousand times rarer. One just has to think of the commotion caused by the discovery by Mary Leakey's team of the ancient hominid prints at Laetoli in Tanzania to appreciate the significance of such an occurrence. Just imagine it. Someone walks across wet sand all those millennia ago, and tracks are somehow preserved for us to see today. Footprints from 100,000 years ago may not sound as dramatic as prints that are 3.5 million years old, but the time period that John had mentioned fell bang into the middle of the emergence of anatomically modern humans. In my gut I felt, even before he had finished speaking, a direct line connected those prints.

"You've got to tell me more," I said. "Who's working on them?" —meaning, who are the paleoanthropologists describing them.

Parkington smiled and drew his chair closer to mine. "Nobody," he said with a grin and glint in his eye. "Why? Do you think you would be interested?"

He then told me the remarkable story of how Dave Roberts, a geologist with the Council for Geoscience, had come across this rarity. The next day I contacted Dave to arrange a visit to the footprints. He seemed genuinely surprised at my interest in them.

The footprints of "Eve" in the sandstone of a fossil dune at Langebaan Lagoon have been dated to approximately 117,000 years before present.

Like John Parkington, Dave was a Scotsman transplanted to South Africa several decades before. He gave the physical impression of being extremely tough, despite his short and wiry frame. As soon as he opened his mouth, his good-natured, joking personality revealed itself, and he regaled with animation his discovery of the footprints.

Dave had been doing a general survey of the sand dunes, studying the unusual geological outcrops and generally mapping the extent of the fossil and modern dunes of the West Coast. On one of these expeditions along the shoreline of the lagoon, Dave noticed some unusual depressions in the sandstone of an eroding layer of fossil dune. On closer inspection he noticed that these were in fact the tracks of a small antelope, clearly impressed in the rock. Intrigued by this find, he worked his way along the beach until he found a second set of tracks, this time they were larger and possessed five toes, clearly those of a medium-size carnivore such as a hyena or wild dog. As he meandered along the beach he chanced upon a couple of crude stone tools manufactured from calcrete, cemented layers of calcium carbonate, that were lying on the beach sand.

Now, if an archaeologist or paleoanthropologist had found these tools, they probably would have examined them and thrown them back onto the ground. Artifacts like these are commonplace, scattered across the countryside in the millions. But Dave was neither. He told me that he had thought to himself that if there were tools and animal prints in these dunes, why should there not be human footprints? Fossil human footprints are so rare to the point of exception, with only five such trackways known in all of Africa. But the find of the tools renewed Dave's enthusiasm for the hunt, and he scoured the outcrops farther down the beach. Believe it or not, within minutes he found exactly what he had been looking for.

Dave took me to a rock located on a well-used public part of the Langebaan Lagoon beach. It was a large, flat piece of dislodged sandstone about two-by-three yards across, tilting precariously toward the water, which was barely 20 feet away. At the base of the rock, people had inscribed graffiti into the soft sandstone. Wonderful, I thought to myself as we climbed up onto the sand-covered surface of the rock. Isn't it just typical of a modern *Homo sapiens* to try and immortalize himself by scratching inanities into what is potentially one of the world's most precious geological rarities. Dave told me that he had covered the prints with sand to protect them from such vandalism.

Dave and I began sweeping the sand off the flat rock with a brush, and I looked critically at the emerging depressions. I was preparing myself for disappointment, because usually such reports of human footprints are false. Either they're pseudo-fossils, impressions that just happen to look like prints but in reality are manufactured by nature, or they are the tracks of some other animal that marginally have the appearance of human prints. But Dave had certainly been right about the two other trackways that he'd shown me earlier, back down the beach. They were indeed the tracks of an antelope and a carnivore, almost certainly a hyena. As I finished brushing away the sand, Dave explained that there were three prints, all pointing downslope. The bowl-like depression of the first print emerged from under my brush. I looked more closely. It was possible that this could be a heel print, and the radiating bulge of sand around the outside of the depression confirmed that these were indeed footprints of something. But were they human? I stopped to examine the depression more closely.

Dave continued the brushing. As he moved farther down, the depression sloped up as one would expect in the arch of a foot. So far, so good. The critical part, however, would be next, and that would be the toes, which are so characteristic of humans that it would decide the issue one way or another. As Dave finished cleaning the first print I leaned over the depression, shifting my angle slightly so that the morning sun would create shadows in the print, highlighting the subtle features. I ran my fingers along the top of the print: one large depression, slightly splayed from the others, a ridge then a small elongate depression, another ridge, a depression, a ridge, a depression, another ridge and the smallest depression. Just above these was the deep rounded bowl of the ball of the foot.

"Well?" Dave asked, sitting next to me with a concerned look on his face.

"It's wonderful," I replied, flushed with excitement and still touching the front of the print. "And it's human!"

In all there were three prints on that rock. He explained that from a geological perspective, based on the position of the sediments in the dune and their height above present sea level, they must be about a 120,000 years old. When I heard him mention that date, which coincided with the appearance of modern human anatomy in Africa, I became very excited. But my enthusiasm was tempered by the chilling realization that something so valuable could vanish before it could be properly studied. We had a pressing problem. We needed to do something fast or these prints were in danger of being washed into the lagoon or, worse, vandalized by visitors to the beach. Casting would be the first step so that, should anything happen to the prints, at least we would have an accurate record from which to work. But this was easier said than done. Bringing an experienced caster and the materials to work such a large, complex surface would not be easy.

With the assistance of a grant from the National Geographic Society, we had a copy of the footprints made within two months, and had begun the process of analyzing the trackway and dating the rock in absolute terms. John Vogel and Stephan Woodborne of the Council for Scientific and Industrial Research once again came to our aid and established that the age of the rock in which the footprints formed was around 120,000 years old. They said there was a possible 8 percent error in their estimate, which meant that the prints were between 130,000 and

110,000 years old. By calculating the position of the sea at the time the prints were made, we refined this date to a best guess of 117,000 years before present. By coincidence, the sea level at that time was just a yard or so higher than the sea level of today. This gives us a real feeling for what the footprint maker would have seen as he or she walked along the beach. It would not have been that different from the landscape today, with the blue waters of the lagoon stretching out into the distance, edged by the granite outcroppings of ancient bedrock ringed by the fynbos that is still evident today.

Eventually, the whole block of prints was cut out of the rock in an elaborate operation involving a helicopter airlift, and moved to the safety of the South African Museum in Cape Town.

It was a miracle the prints were preserved at all. By examining the geology of the rock and the surrounding dune, we could reconstruct the events leading up to the formation of the prints 117,000 years ago. It would appear that a heavy rainstorm saturated the dunes on the lagoon shore. Shortly thereafter a person crested the dune, walking from the direction of the ocean-facing beach west of the lagoon. This individual was short, probably around 4.5 feet tall and barefoot. We'll never really know whether that dune walker was a small man or a woman, or even a child, but judging by the size of the prints—which are only 8.5 inches in length—it was probably a small individual. We can surmise that this person was alone as she made her way down the north face of the dune, meandering between the dense bush toward the beach. What we do know is that just above the high-water mark and still on the dune slope, she took three steps that pressed deep into the sand, spreading small concentric rings away from her print.

Stepping off the slope of the dune, she walked onto the harder sand of the beach, heading eastward along the shoreline. Within a short time, the winds picked up from the southwestward ocean side of the dune. As the wind raced across the back of the dune, it began to dry the top of the dune through evaporation, but the footprints lying on the northern aspect of the dune were protected from the wind, maintaining the moisture that preserved the shape of the prints. Within minutes slumping began to occur. Mini avalanches caused by the dry sand at the top of the dune cascading down onto the still-wet sediments near the waterline began to increase in frequency. Soon a small wave of dry sand about five inches in depth covered the footprints, filling the still-wet impressions with dry sand and burying them under a thin protective sheet.

This burial alone, however, would not have been enough to preserve the footprints for over 100,000 years. The very nature of the dune sand was the key to the fossilization. The sand was comprised mainly of silica and small fragments of seashells. This calcium-rich sediment, once bound by the precipitation of carbonate, formed the concrete-like sandstone that preserved the prints for over 110 millennia.

Who was the maker of the footprint trail? Was she one of our species? The fossil evidence for modern human origins in the critical time period between 100,000 and 200,000 years ago is, like that of the archaics, frustratingly rare. In fact, the entire fossil evidence of anatomically modern-looking humans from around the world, and over 60,000 years, wouldn't fill a card table. Nevertheless, the fossil evidence that does exist points to an African origin for modern humans.

Specimens representing anatomically modern humans have been found across Africa, but the best evidence consistently arises in southern Africa. From the Omo Basin in Ethiopia is a skull, clearly modern in its morphology. Uranium-series dating on shells found in the same sedimentary layer as this specimen provide a tentative date of around 130,000 years, but there is the possibility of intrusion into the sediments where the specimen was recovered. More reliable are specimens from South Africa originating from two sites, Klasies River Mouth and Border Cave. Both sites have yielded remains of individuals that must be considered modern anatomically. The Klasies River Mouth remains have been dated by a variety of methods to around 100,000 years before present. The Border Cave specimens, including the skeleton of the buried child, date to around 80,000 years old. As you move northward into the Near East, we see another very early modern *Homo sapiens* cranium at the Quafzeh cave in Israel. This tantalizing material, comprising a skullcap of a female, appears modern and dates to possibly 90,000 years old. But once outside of Africa, there is little fossil evidence of any candidate for truly modern human morphology anywhere near the ages of these finds. Thus, the fossil evidence leads us to the conclusion that modern human morphology comes out of Africa.

DNA also comes to the support of an out-of-Africa origin for our species. Genetic evidence from mitochondrial deoxyribonucleic acid (mtDNA) suggests that we humans share an unparalleled (among the ape community) genetic unity. MtDNA are tiny sausage-shape structures that live inside our cells. They convert energy by changing basic substances into chemicals that our cells can more readily use. DNA is inherited by children from their parents, but mtDNA is only inherited from the mother. In the 1980s, scientists at the University of California, Berkeley, were the first to suggest that mtDNA could be used to clock the evolution of our species. Allan Wilson, Mark Stoneking, and Rebecca Cann sampled mtDNA from women representing different racial groups and peoples living in different geographical regions around the world. Their simple goal at the time was to find out how much mtDNA differed among these populations. Theoretically, the similarities and differences could give some indication of how closely or distantly the sample populations were from each other. Quite simply, the closer the mtDNA, the more closely related the group. An offshoot of this work is the hypothesis that mutations in mtDNA occur at a relatively fixed rate. By calculating the rate of mutation one should be able to date the last common ancestor of all the populations studied.

The results of Wilson, Stoneking, and Cann's research, as well as subsequent DNA research conducted around the world, has reached the remarkable results mentioned above, namely a common origin of all living humans as recent as 100,000 years ago. Furthermore, African peoples show a greater diversity in their mtDNA than do any other groups from around the world. This suggests that modern humans have been living in Africa longer than anywhere else. The genetic-Eve hypothesis is just a logical extension of these results—that is, that all modern humans can trace their origins to a small group of females, in fact a single female, living somewhere in Africa between 100,000 and 200,000 years ago.

The University of the Witwatersrand, through the international human genome project, has begun collecting genetic data from different ethnic groupings to try and map the sequence of human development, which may cast additional light on whether the Khoisan are the original ancestral population. Geneticists Trefor Jenkins and Himla Soodyall have convincing evidence that the Khoisan, Negroid, and Pygmy populations of Africa have different evolutionary histories, having separated from a common stem around 150,000 years ago. This is by no means conclusive and their findings will undoubtedly be modified as the project unfolds.

One of the firm believers in the theory that modern humans arose in South Africa is Hilary Deacon, professor of archaeology at Stellenbosch University in the Cape. Through his excavation of sites like Klasies River Mouth, he has made a convincing case that the Khoisan people of the Western Cape are the last vestige of a living link to the stone ages. Known before the era of political correctness as the Bushmen, their ancestors have been resident in the subcontinent for tens of thousands of years.

These were the people responsible for the first art in the world, probably the first burials, the use of ocher, and the development of other culturally modern forms of behavior.

So there is clear potential that southern Africa is the source area for modern humans, providing evidence in the form of fossils, culture, and even our own genes. But there is also further evidence. The geographical situation is unique within Africa. The central regions are dry and filled with deserts and extensive grasslands. The mountain ranges that divide this inland from the coastal area provide a potential isolating barrier for populations existing inland and on the coast. Additionally, we have in the southern African Cape coastal region a uniquely diverse plant kingdom, the fynbos. This, one of only seven plant kingdoms on the planet, is confined to the narrow strip of coast between the Cape coast and the inland mountains. An entire plant kingdom in a tiny geographical area. Humans caught between the ocean and the inland mountains would have been forced to exist in this unique ecological zone, developing new strategies for survival that no other populations in Africa would ever be forced to encounter. That the earliest definitive fossil evidence for the emergence of modern humans comes from this very same region—and that there is emerging evidence that the earliest appearance of modern human behavior also comes from this narrow coastal region of Africa—may not be a coincidence.

South Africa is a country that is familiar with the concept of isolation. Besides it's geographical relationship to the rest of the world, the past 40 years have been a period of political isolation because of the politics of apartheid. Yet isolation may have been the trigger for producing these anatomical moderns. Cut off from the rest of Africa by the arid conditions of the hinterland, an isolated population eking out an existence on the Cape coastal plains began to take on a modern morphology, led mostly by a physical gracilization and an increase in brain size. The Out of Africa hypothesis that has been put forward by scientists like myself and

Christopher Stringer of the Natural History Museum of London holds that this genetically modern population eventually expanded out of the region, through Africa and into the rest of the world. Their superior brainpower and adaptive cultural behavior gave them the edge over the *Homo heidelbergensis* and *neanderthal* populations that were the descendants of the first wave of expansion of *Homo erectus* out of Africa around 1.5 million years ago. Within the relatively short space of a few tens of thousands of years, these relative newcomers replaced the existing *Homo* populations that had scattered around the world. This argument, which implies a relatively recent origin of modern humans, is not without its critics. The adherents of the so-called multiregional hypothesis believe that the origins of different peoples in the world are far deeper—the roots being in regional adaptations that evolved out of the first hominid migration out of Africa. This debate has not yet been resolved completely and is likely to dominate the question of modern human origins for some time yet. But with the emergence of genetics into the study of human origins, the answer is not far off, and all indications point to Africa as the winner of this debate.

What is clear today is that from around 130,000 years ago there is a cultural shift among hominids, the evidence of which runs along almost the entire southern African coastline. On the fossil evidence, there is a clear case to be made that anatomically modern humans did emerge from this region. If they simultaneously appeared in other parts of the world, the evidence for this has not yet been found. My hunch is that South Africa is where it began, and that among this original population of moderns was a woman who carried the genetic potential of all humans yet to be born. I believe that it is essentially her genes that we still carry with us today, whether we come from Chad or China, Arkansas or Armenia. In a sense, it is her footprints that have been left behind on the sandy slopes of Langebaan Lagoon 117,000 years ago.

CHAPTER 13

FINAL STEPS

~

There are sacred places on this Earth: sites that are transcendental in that they resonate with meaning not purely of our making. I don't offer this in a conventional spiritual or religious sense, because my science makes me skeptical when it comes to issues of pure faith. All too often the name of God is invoked by way of explanation for things we cannot understand or do not have the courage to question. This is not to say that we have all the answers, or even the potential to demystify the workings of an infinitely complex set of chemical relationships that we call the universe. What I mean is that there are pockets in the landscape that have become repositories of human consciousness if one is aware of their history, where not even the overwhelming passage of time can obliterate the residue of our birth as a genus, as a species. There are many such places in Africa. It has been my privilege to explore a lot of these sites as I attempt to understand who we are and from where we have come. In a way, I think of this as a journey in Eve's footsteps, tracing the path walked by our ancestors.

I sometimes find myself sitting on the crest of a sand dune at Langebaan Lagoon, staring out to sea and listening to the kelp gulls and sandpipers that flap past in the dizzy gusts of wind whipping along the Cape West Coast. Sunset is a particularly poignant time to contemplate the view across the fynbos, toward the other end of the lagoon, where the domed hills jut out against the skyline. Besides a moderately shifting sea level, this landscape essentially would have been the same environment that our most immediate ancestors lived in during the past 200,000 years. Today there is still little evidence of human habitation from this vantage point, so it is not difficult to cast oneself back in time to contemplate the life and times of these first anatomically modern fledglings. Burnished by the sun and the sea air, surviving on a diet of seafood, plant life, and game, their actions formed the blueprint of what we are today. Their imprint is still with us, stamped in the genetic footprints of mitochondria that link our morphologies with the birth of our species.

What have we learned so far? Quite a lot in general about human evolution, a little less about the specifics than we would like to know. What is crystal clear is that in the short space of 75 years, we have confirmed that Africa is, in the greater sense, our birthplace. It is our cradle and our guidepost to the future. I am awed by the richness and complexity of humanity's long journey and continually amazed by our young science. We have probably added more knowledge to our understanding of human origins in the last 20 years than in the 200 years that have elapsed since Chevalier de Lamarck's intellectual breakthrough, when he realized that nature is a dynamic process in which living organisms change according to their environments. Lamarck has been lambasted for his misconceptions about the way inheritance occurs, but his *Philosophie Zoologique* published in 1809, inspired by his study of the humble mollusk, was the first major work suggesting that evolutionary lineages occur in the natural world. Struggling to free himself from the shackles of biblical convention, Lamarck was prepared to countenance the possibility that humans had arisen from an ape-like animal.

It was Darwin who said that this event probably occurred in Africa.

Unbelievably, almost 150 years since Darwin's *Origin of the Species*, the chains of creationism are still strongly evident, even in the most scientifically progressive societies like the United States. For me, it is deeply disturbing that the Kansas Board of Education voted in 1999 to remove evolution from the state's science curriculum, even more so since I happen to have been born in that state. I think our real challenge is the continual effort that is required to free ourselves from the constraints of the science that has gone before us, to ensure an open mind in accepting new ideas even if they impinge on the comfort zones of our own understanding.

In the study of evolution, it is intellectually suicidal to announce that one has reached an endpoint in comprehending the origin of our species. The last decade alone has changed so much of our understanding of human origins; our knowledge of individual species in our family tree and the recognition of the very diversity of that tree have grown in leaps and bounds. At the beginning of the 1990s, there were about eight species of hominid known to have existed between five million years and one million years ago. Today, we recognize no less than 13 early hominid species, and the numbers of "new" species will grow as the science progresses.

From my sand-dune perspective on the Cape coastline, one of the more fascinating developments during the past ten years has been the increasing body of evidence that points to the origin of anatomically modern humans from this region. Every time I dig my toes into the white sand at Langebaan, I visualize what it must have been like for those early humans who survived by scanning the shoreline for shellfish like the mollusks studied by Lamarck. Another significant development in my view has been the acceptance of how much hominid diversity there has been in the last several million years. But it is one thing to recognize this diversity, and it is another matter entirely to try and make sense of it. The hominid fossil offerings from the ancient sediments of Africa have complicated our understanding of humanity, over-turning the simple hierarchical model spawned by Lucy's discovery. This has had the effect of covering in sand Eve's footsteps from her own ancestors, and the debate about the missing link—a misnomer of a concept if ever there was one—appears to be as intense today as it was in the era of Dart and Broom. That's because of all the new pretenders to the throne.

One of the more recent players is a new ape-man species called *Australopithecus gahri*, a name meaning "surprise." But is it a surprise? *Gahri*, from Ethiopia, dates to about 2.5 million years. That's an important date in the evolutionary calendar because it is the theoretical terminal-end of *africanus*'s reign, the beginning of the emergence of the robust australopithecine experiment, and it marks the doorstep for the emergence of the genus *Homo*. Does this species, as has been suggested, fulfill the criteria of being a missing link between the gracile australopithecines and *Homo*?

To argue as Tim White seems inclined—that *gahri* fills the gap between *afarensis* and our own genus *Homo*—I believe is both contradictory and a gross oversimplification of the evidence. *Ghari* shares *africanus*'s dental problem in that its big back teeth represent either a backward step in human evolution from Lucy, or it is evolving in a different direction toward the robusts. It is hard to see how it would then lead to *Homo*. But as I hope I have emphasized, evolution is not all in the head. *Garhi* apparently has other interesting traits in its body shape that may include or exclude it from the possibility of being the ancestor of *Homo*. It seems to have very long forearms and long legs. A lot is going to depend on what the bodies of these earliest members of the genus *Homo* look like from the neck down. We don't really know yet what the body shape is of *Homo habilis* or *Homo rudolfensis*; more complete specimens have to be found before we can start answering that question.

Our knowledge of early hominids has also expanded with the spread of the known geographical range in which their fossils have been found. In the 1980s, the distribution of these bipedal creatures was confined to Tanzania, Ethiopia, Kenya, and South Africa. Now we find the gap split between East and southern Africa by the Malawi hominids, while there's also been a westward expansion of the evidence to Chad with Michelle Brunet's new species *Australopithecus barhagazelli*. This is another gracile australopithecine much like *afarensis* in its morphology and dating to around three million years. And of course, the "old" species like *africanus* are beginning to have new importance in our understanding of just how diverse and wide the branches of the human family tree have been. Nonetheless, I still think there is a tendency to oversimplify the problem of our origins. We carry within us the need to recognize each and every fossil we find as a direct ancestor, but it is clear that the whole picture is not yet visible. Each species of early hominid represents merely a pattern, a symbol of what was going on at any one time or place in the past.

I began this book as a focused attempt to redress an imbalance in the attention given to South African discoveries relative to those of East Africa. The South African contribution has been sidelined if not ignored, because of the nature of funding and the publicity surrounding significant finds as well as the vested interests that have dominated East African paleontology for the past half century. Despite the renaissance that the search for human origins in South Africa has enjoyed over the past five years, there is still a lingering bias against this region. In a *Time* magazine article from August 1999, for instance, the idea of an East Side evolution for humans was clearly supported above all others. The entire South African record, which consists of at least 40 percent of the whole early hominid record from Africa, was relegated to two small captioned boxes. A fuller examination of the evidence needs to occur if we are going to reach an understanding of human evolution.

Although I have argued the case for a southern African origin of humans, I believe that the underlying message that arises from a dispassionate view of the present evidence is that we are not in a position to identify a single region of Africa as the birthplace of humankind, particularly in the early hominid record. If anything, the evidence at hand suggests many evolutionary experiments were conducted during the course of the last four or five million years of human evolution across the continent. And there is certainly more to be discovered before we understand even single time periods in Africa's past.

It may seem shocking to say, but there is the very real possibility that not one single fossil that we have found so far either in East, South, or northwest Africa is the direct ancestor of living humans. These early and later hominids do represent the general pattern of human evolution over the past five million years, but the missing link might still be missing. Is there more to know, more to find? Of course there is. We stand only on the tip of the iceberg, and yet we know more about human evolution now than ever before. We also know more about what we haven't found yet, where the real missing links are. In the 75 years since the announcement of the Taung Child discovery, we have come a long way in understanding human origins, but the most thrilling aspect of this knowledge is to realize just how much more there is to learn, discover, and understand.

I have clearly and strongly supported the argument for *africanus* as the potential ancestor of the *Homo* lineage, but is there enough evidence at hand to definitively make this claim? The few scraps of evidence we have of the postcranial morphology of the earliest members of the genus *Homo* seem to indicate that these ape-men had long arms and short legs, and that evidence alone seems to make *africanus* a better ancestor for early *Homo* than the more human-proportioned *afarensis*. But to base such an assumption on maybe two or three fragmentary specimens is probably squeezing more from the evidence than one should. What I am convinced of, however, is that human evolution between three and two million years ago is far more complex than we had previously suspected. To continue to draw ladders where there are bushes is to ignore the evidence.

The overriding lesson for me is that Africa has taught us as a species to be adaptable. It is the stress of encountering diversity in Africa that has given us extraordinary abilities to adapt to change. Throughout our evolution, this constant encounter with African diversity has given us extraordinary adaptive abilities. In the last hundred thousand years alone, we as a species have not only been able to dominate all of the land of this planet but marine environments as well, and we have extended our range beyond the biological sphere of Earth into space. That is quite an achievement for a little African ape.

Many readers may view the various species of early hominids explored in this book as failures; after all, we're here and they're not. But each of these species existed at least 300,000 years, and some almost a million years—between three and ten times longer than modern humans have lived on the Earth. Given the rapid capacity we have developed to destroy ourselves, there are no guarantees that *Homo sapiens sapiens* will survive to reach the temporal range of even the greatest failures among our family tree. But in knowing our history, and our *own* history includes these African ape-men, we gain a window into our future. Africa guides us toward that future. It is the proving ground of our species and of our kind. It always has been and always will be. The roots of our family tree are undeniably and firmly planted in the soils of Africa. Every critical event that has molded us physically, technologically, and spiritually has an African origin. Bipedalism, enlargement of the brain, tool use, fire, the modern human mind and body are all African inventions. It is thus also our time machine to the future. For what is happening in Africa today is a signpost for the future of the human race. The disease, famine, infant mortality, warfare, and genocide that characterize

Africa for many are only the low points. The experiments in human understanding and democracy, the successful interactions of the First World and Third World, and the struggles to conquer disease and poverty that are taking place today in Africa are the hope for our species. If we confront and fix the problems that face our species in Africa, then we fix them for the rest of humanity. In the long run, we can do more for our species by simply recognizing that we *are* fundamentally an African animal.

As Africa holds the key to our origins, it also may hold the key to our extinction. The very disease that currently threatens the survival of our species more than any other originated in Africa. The source of AIDS has been traced to chimpanzees living in equatorial Africa. HIV1, the type of human immunodeficiency virus that causes most forms of AIDS, is a variation of the rare simian immunodeficiency virus (SIV) that affects monkeys and apes. Because one of the most common ways of transmitting the virus is through blood, it is probable that the disease spread to humans as a result of people hunting chimpanzees for food. Yet chimpanzees may be immune to AIDS; chimpanzee SIV is exceptionally rare, with only four cases recorded by science.

Could there be a clue for the scientists trying to contain the AIDS pandemic in working out why so few chimpanzees get ill from HIV, yet people do? Perhaps. However, with chimpanzees facing extinction because of human hunters and land loss, the answer to our own salvation may be removed by our own hand. Would it not be the most bitter of ironies if our closest cousins in the primate world disappear, and with them goes the clue to our survival as a species? The world ignores Africa at its peril.

As I sit on those Cape dunes, I often wonder whether Eve had contemplated her own origins in the way that I am doing now? Was she capable of such thought? What would have gone through her mind as she watched the waves spill onto the beach and felt the sand between her toes? Was she thinking purely about her next meal, or could she comprehend the infinite beauty of this landscape and communicate these feelings to her closest companions? Did she ponder the possibility that I might sit here today, over a hundred thousand years later, thinking about her? Who will sit here one hundred thousand years in the future and think about both of us?

INDEX

Boldface indicates illustrations.

A

Afar region, Ethiopia: hominid fossils 32, 112–113

Africa: biodiversity 9, 20; birthplace of humans 6–9, 13, 18, 21, 75, 103, 140, 299, 304; gap in fossil record 42, 43, 221, 275, 282; largest habitable landmass 8–9, 20

Afrikaners 95, 238; apartheid 61–62, 102–103, 140, 142, 278

Antelope 1–4, 38, 43, 132, 148, 154–155, 157, 160, 261, 282–283, 293–294

Ape-men 13, 15, 19, 36, 42–43, 46, 87, 105, 151–153, 188, 303, 307; see also Australopithecines

Apes 19, 24, 67, 79; evolution 20–21, 25, 220–221; see also Ape-men; Australopithecines; Chimpanzees; Gorillas

Ardipithecus ramidus 31–33, 201, 203–204, 214, 230; bipedalism 32, 221, 222; fossil 32; name origin 32; oldest known hominid 32, 221; size 32, 33

Art, prehistoric 10, 17, 96, 300

Asfaw, Berhane 203, 206, 225

Australopithecines 15, 31, 90, 97, 101, 105, 107, 109, 154; bipedalism 26, 31, 92, 94, 166–168; brain 31, 92; extinction 31–32, 265; fossils 96, 104, 105; predation by carnivores 151–152, 153–155, 163; teeth 149–150; time on Earth 265; see also Gracile; Robust; individual species by name

Australopithecus aethiopicus 36, 38, 119, 224

Australopithecus afarensis 33, 36–37, 45, 47, 49, 117–119, 168–171, 214, **219,** 224; as tree-climber 166–167; bipedalism 166–168, 170, 223; body proportions compared with those of africanus 34, 47, 165–166, 172, 174, 177–191, 194–195, 199–202, **203,** 205–207, 212, 214–218, 220, 224, 246–247, 257, 267; name origin 33; sexual dimorphism 33, 34, 175, 223; size 33; skull 34; teeth 33, 37; see also Lucy

Australopithecus africanus 32, 37, 39, 45–47, 77, 87, 118–119, 150–151, 164–177, 188–189, 214–215, 217, **219,** 239, 241, 306; arboreal lifestyle 34, 217, 223; as original hominid ancestor 49–50, 53, 307; body proportions compared with those of *afarensis* 34, 47, 165–166, 172, 174, 177–191, 194–195, 199–202, **203,** 205–207, 212, 214–218, 220, 224, 246–247, 257, 267; emergence in South Africa 34, 94; missing link theory 46; teeth 36, 47, 169, 267; *see also* Stw 431; Taung Child

Australopithecus anamensis 191, 193, 199, 214, 220; bipedalism 32; fossil remains 32–33, 222; teeth 195; weight 33

Australopithecus bahrelghazali 33–34, 215, 306; fossil skull 33

Australopithecus boisei 38, 108, 169, 224, 267; cast of skull **273**

Australopithecus ghari 36, 38, 225, 305; stone tools 37

Australopithecus prometheus 94, 151, 264

Australopithecus robustus *see* Robust australopithecines

B

Baboons 261; diet 272; fossils 71, 72, 94, 140–141; giant 282; prey of black eagles 160, 161, 162; skulls **160,** 162

Berg Aukus femur 288

Berger, Lee **236**

Bipedalism 8, 10, 17, 19, 25–32, 46, 57, 74, 92, 94, 99, 113–116, 166–168, 170, 175, 194, 198, 200–223, 229

Black eagles 159–161; baboon skull from nest **160**; bone accumulations 160; killing of vervet monkey 158–159

Black Skull (KNM WT 17000) 36, 38, 118–119, 224

Bodo, Ethiopia: fossil skull 188

Border Cave, Natal, South Africa: hominid burial 10, 290–291, 298

Bouri, Ethiopia: hominid fossils 36, 37

Brain, Bob 41, 128, 129, 151–156, 163, 227–228, 268–270, 272

Breccia 72–73, 86–87, 92, 99, 128, 133, 148–149, 240, 273, 284

Breuil, Abbé Henri 95, 96

Broken Hill Man 286–287, 288; skull 43, 287

Broom, Robert 80, 83–89, 95, 99, 131, 150, 174, 238, 267, 274, 280; awards and recognition 92, 98, 101; bust of 248; death 100, 101; loss of governmental support 102; monograph 92, 93, 100, 145; personality 84, 88, 90, 97, 98

Burials 10, 17, 125, 151, 155, 300

Buxton Lime Quarry, Buxton, South Africa: fossil skulls 71–73, 140

C

Chad: hominid fossils 105, 215, 306

Chimpanzees 8, 18, 19, 26, 104, 135, 186, 223, 279; abstract thought 137–138; aggression 136–137; behavioral studies 135–138; captive 13, 135; evolution 25, 221; feeding 26, 135, 137; incest taboo 139; jaws 74; limbs 186, 187; mirror experiments 137–138; sexual dimorphism 175, 223; skeletons **203,** 204; social groups 224; source of AIDS virus 309; subspecies 135; tools 137, 264, 270, 271

Choukoutien Caves, China: Peking Man 81, 153; use of fire by hominids 153

Clarke, Ron 162, 183–184, 227–229, 231–234, 239–241, 267, 273

Climate: aridification 9, 24, 56, 267, 272; changes in 9, 10, 23, 25, 31, 34–35, 132, 267, 271; *see also* Ice ages

Coppens, Yvves 104, 105, 110, 112–113

D

Dart, Raymond Arthur 34, 63, 67–74, **75,** 76–78, 85–86, 91, 95, 125, 151, 156, 226, 238, 264; aggression theory 151–154; criticisms of 80–83, 89; emotional problems 71, 81, 88–89; Makapansgat cave excavation 94, 175; scientific papers 75–76; Taung skull 69, 70–77, 140, 152, 164; vindication of theories 92–93, 98; wife 73, 81, 88

Darwin, Charles 96, 107, 278–279, 280

Dating methods 130; aluminum 134; argon-argon 131; chlorine 36, 134; comparative faunal 131–132, 134, 202; Electron Spin Resolution (ESR) 134; geochronology 286; paleomagnetics 132–133; potassium-argon 131, 170; uranium-series 297; x-ray defraction 132

Dioxyribonucleic acid (DNA) 6, 298, 299

Drimolen, South Africa 226, 257, 273; bone tools 268, 270; excavation techniques 241; fossil discoveries 14, 235, 236, 243, 274

Dual Congress, Sun City, South Africa 213–214, 228–229, 243, 245

E

Eagles 158–159, 160–162; hominid predation 163; killing of children 162

Ecosystems: forests and woodlands 9, 19, 21, 23–26, 32–34, 36, 38, 40, 135, 221, 283; fynbos 7–8, 296, 300, 303; savanna 24–26, 28, 35, 38, 41, 56, 268, 271, 283

Elandsfontein, South Africa: fossils and stone artifacts 280–281

Eve (progenitor) 2–6, 11–13, 17–18, 24, 30, 288, 299, 302, 309–310; footprints 5, 6, 12–13, **292,** 296–297

Evolution: linear model **202–203;** opposition to theory of 82–83, 95, 102, 103, 238; Out of Africa theory 6, 7, 300–301; pulse theory 22–23, 24, 34, 36; survival strategies 35, 134

Extinctions 272–273; dinosaurs 19, 272; hominids 31–32, 39, 41, 47, 265, 268, 271, 274–275

F

Fire-making 8, 41, 151, 152–153, 308

First Family (hominid bone collection) 114, 171, 194

Food gathering 23, 26–28, 43, 55–56, 139, 259–262, 266, 268

Footprints, fossil 4–6, 11–13, 24, 53, 115–116, 227, 291, **292,** 293–297, 301–302

Fossils: accumulating agents 127, 141, 151–154, 155–157, 159–160, 238, 269, 286; fossilization process 127, 128, 154, 297; gaps in record 42, 43, 221, 275, 282

G

Genetics 6, 49, 298, 299, 301

Gladysvale, South Africa: aerial view **143;** baboon skull **160;** hominid fossils 13–14, 66, 142–143, 146–150, 155–156, 227, 236, 241, 246, 257; hominid teeth **66,** 149; saber-tooth cat skull 143–144, 147

Goodall, Jane 135–138

Gorillas 104, 175, 221, 223, 279

Gracile australopithecines 31, 34, 36, 38, 94, 99, 109, 149–151, 265

Great Rift Valley, Africa-Asia 13, 103–104; fossils 104, 105

Gurche, John 203, 213, 216–219

H

Hadar, Ethiopia: hominid fossils 37, 47, 53, 112–114, 117, 126, 168, 195, 208–211

Hoedjiespunt Child 284–286, 290, 291

Hominids: aggression theory 151–153; ancestral apes 12, 25, 30; bipedalism 8, 10, 17, 19, 25–30, 32, 46, 57, 114–115, 308; brain development 28–29, 35, 123, 300, 308; categories 31; evolution 13, 18, 20, 28–30, 36,

46, **202–203,** 221; migrations out of Africa 41; name origin 19; new species 107; number of species 29, 304; oldest known 32, 221; postcranial anatomy 15, 49, 109, 126, 165, 203–204, 217, 220, 235; skulls 49, 123, 143; survival strategies 35, 134; teeth 59, **66,** 149, 271, 284–286

Homo (genus) 17, 19, 28, 30, 32, 34, 37, 46–48, 53, 214, 224–225, 243, 265, 274; ancestry 47, 49, 50, 105; cranial capacity 38, 39, 109, 271; first known species 38; meat-eating 271, 272; new species 109; split from chimpanzee lineage 18, 32, 135; survival strategies 35

Homo erectus 39–41, 46, 106, 109, 111, 117–118, 169, 225, 227, 275, 287–288, 290; fossils 41–43, 152; hunting 40, 272; increase in brain size 43, 275; migrations 41, 275, 301; teeth 43; tools 40, 42, 275, 289; use of fire 41; see also Java Man; Peking Man; Turkana Boy

Homo ergaster 39, 169, 239, 287, 289

Homo habilis 38, 45, 63, 108–112, 119, 126, 169, 190, 224–225, 239–240; extinction 39; skull 124, 236

Homo neanderthal 42, 46, 78, 79, 301

Homo rhodesiensis 43, 286, 287–288

Homo rudolfensis 38, 39, 169, 305

Homo sapiens 8–9, 11, 15, 17, 26, 43–44, 46, 169, 286–288, 298; population dispersals 9; tools 9–10, 289

Homo sapiens (archaic) 43, 286, 287, 290, 297

Homo sapiens sapiens 18, 19, 46, 265, 308

Hughes, Alun 166, 185, 199, 229, 238–239

Humans, modern 16; African ancestry 21; lineage shared with chimpanzees 8, 18–19; skeleton **203**

Hunting 40–41, 42–44, 272

Hyenas 127, 141, 148, 262, 269, 293; predation of hominids 152–153, 156, 163, 264, 284–286

Hyraxes 155, 156, 160; prey of black eagles 159, 162; skulls 160, 162

I

Ice ages 9, 23, 34–35, 104, 280

J

Java Man 82

Johanson, Donald 37, 46–47, 49–50, 64, 67, 99, 112–115, 118–119, 164, 169–170, 178, 190; discovery of Lucy skeleton 33; rivalry with Richard Leakey 52, 53, 54

K

Keith, Sir Arthur 76–77, 78–80

Keyser, Andre W. 235–236, **236**, 237, 243, 262–263, 268, 272, **273**

Kimeu, Kamoya 117–118, 192

Kitching, James 94, 175, 238, 239, 280

Klasies River Mouth (site), South Africa 10, 299; fossils 298

KNM-ER 1470 (hominid) 109–110, 112, 126; brain capacity 112

KNM-ER 3735 (hominid) 170, 190, 197, 200

Koobi Fora, Kenya 120, 126; hominid fossils 55–60, 111, 170, 196

Kromdraai, South Africa 156, 257, 267; hominid fossils 39, 90–92, 95, 97, 103, 141, 274

L

Laetoli (region), Tanzania: hominid footprints 53, 115–117, 227

Langebaan Lagoon, South Africa 1–5, 8, 11, 15, 280, 293–294, 301–302; fossils 5, 6, 12–13, **292,** 296–297

Language and speech 18, 29, 40, 44, 112, 290

Latimer, Bruce 203, 206, 208, 209, 210

Le Gros Clark, Sir Wilfred 77, 95–96, 98, 109

Leakey, Louis S. B. 38, 48, 52–53, 56, 63, 95–96, 106–112, 120–123, 135, 184, 224

Leakey, Mary D. 48, 53–54, 63, 95, 106–109, 115–117, 120–121, 124, 227, 228

Leakey, Meave 33, 106, 191–200, 204, 206, 214, 222, 228–229, 274

Leakey, Richard E. 39, 47–49, 58, 60, 106, 110, 112, 115, 117–118, 123–124, 199–200; airplane crash 111, 120, 192; campaign against ivory trading 111, 120

Leopards 127, 142, 148, 156; hominid predation 142, 151–155, 163, 258, 260–261, 272

Linnaeus, Carolus 19–20

Little Foot (hominid foot bones)

229–231, 249–250, 252

Long arms—short legs theory 177–187, 189, 191, 199, 201, 202, 212–219, 247

Lovejoy, Owen 27, 168, 203, 205–206, 208

Lucy (australopithecine) 45–50, 52–54, 113, 116, 165, 170–176, 186, 196–197, 202, 205, 208, 305; bipedalism 114, 168, 223; brain capacity 114; evolutionary role 47–49, 115, 201, 205; height 117; skeleton 33, 47, 53, 67, 124; teeth 125

M

Makapansgat, South Africa 94, 156, 245, 246; animal fossils 141; distribution of large, medium, and small animals 157; hominid fossils 93, 103, 141, 151, 175, 202, 238

McHenry, Henry 212, 213, 214; theory on hominid limbs see Long arms—short legs theory

Miocene epoch 9, 23, 24, 104, 278, 282

Mitochondrial DNA (mtDNA) 6, 298, 299

Monkeys: fossils 140–141, 155; vervet 157–159; see also Apes

Mrs. Ples (Sts 5) 97-98, 140, 164

N

Napier, John 108, 109, 224

Nariokatome, Kenya: fossil discoveries 39, 117

National Geographic Society: funding of hominid research 108, 111, 205, 295

Neanderthals 42, 46, 78, 79

O

Ocher 3, 10, 30, 290–291, 300

OH 62 (hominid) 110, 119–120, 170

Olduvai Gorge, Tanzania 126, 227, 236; hominid fossils 38, 52, 105–107, 110, 170; stone tools 40

Omo Valley, Ethiopia: fossil finds 54, 110–112, 297

Orpheus and Euridyce (austalopithecines): jaws 262–263, 273, **273**

Osteo-Dento-Keratic (ODK) culture 151, 152, 154, 264, 269

P

Paleoanthropology: definition 16; forensics 122–123, 125, 139; hoaxes 78–79, 81–82; see also Dating methods

Pan African Prehistory Conferences 94–96, 102–103, 107–108

Paranthropus (near-man) 31, 91, 267

Parkington, John 280, 283, 284–286, 291–292

Peking Man 81, 82, 153

Piltdown Man 78–79, 81–82

Pleistocene epoch 154, 214, 275, 282

Plesianthropus transvaalensis 98

Pliocene epoch 9, 34, 36, 104, 133, 154, 169, 214–215, 272, 275, 282

R

Roberts, Dave 5, 6, 11, 292–293, 294, 295

Robinson, John T. 97–98, 109, 175

Robust australopithecines 31, 34–36, 38–39, 45, 152, 224, 237, 244, 258–275; as separate genus 267; bipedalism 99; cranial capacity 266, 268; evolutionary lineage 267; extinction 41, 47, 265, 268, 271, 274–275; group structure 259–260; jaws 36, 266, 267, **273**; killed by leopards 258, 260–262, 272; lifespan 266; number of species 267; possible tool use 99, 271; sexual dimorphism 259–260; size 259; skull 259, 262–263, 266; teeth 36, 260, 266–267, 269, 271; time on Earth 265; vegetarian diet 259–262, 266, 268, 271–272

S

Saber-tooth cats 143–144, 147; predation of hominids 126, 153, 163; skulls 143–144, 147, 282

Saldanha Man: fossil 8, 43, 281–282, 288

Smuts, Jan Christian 83, 85, 95, 97, 102

South Africa: as birthplace of human ancestry 280, 299, 300, 301; map of early hominid sites 14; politics 51, 55, 61, 65, 101–103, 140, 239, 277; see also Drimolen; Gladysvale; Langebaan Lagoon; Makapansgat; Sterkfontein; Swartkrans

South African National Museum, Cape Town, South Africa: hominid research 282, 296

Sterkfontein, South Africa 41, 46, 89–95, 103, 130, 133, 146, 156, 165, 174, 199–202, 226, 231–232, 236, 243–245, 258, 267; cave members 35, 129–130, 134, 229–230, 239, 241; distribution of large, medium, and small animals **157**; excavations 232–233, 237, 240–241; hominid fossils 13, 35, 39, 63, 66–67, 86–87, 130, 178,

183–184, 197, 234, 240, 245; monkey fossils 141; speciation events 35, 39; World Heritage Site 67, 248, 257

Stone Age 10, 14, 283, 285; tools 37, 40–41, 134, 275, 280, 283, 289–290

Stw 53 (hominid) 39, 239, 240

Stw 431 (hominid) 165–166, 171–173, 175, 178–179, 193, 215, 218–219, 245; skeleton 166, 171, **172**

Stw 505 (hominid) 196, 218, 219, 242, 244

Stw 517 (hominid) 206–211

Stw 573 (hominid) 67, 126, 256, 257

Swartkrans, South Africa 156, 257, 272; bone tools 268, 269, 270; distribution of large, medium, and small animals **157**; hominid fossils 13, 99, 100, 103, 150, 227–228, 274; monkey fossils 141; use of fire 41, 152–153

T

Tanzania: fossil discoveries 13, 38, 52–53, 170, 306; see also Laetoli; Olduvai Gorge

Taung Child 34, 84–85, 88, 92, 126, 140, 164–165, 228, 307; bipedalism 74, 98; brain 73, 74, 82; cause of death 151, 157, 161, 162; named as new species 74; skull 34, 46, 69–74, **75,** 76–78, 80–81, 83, 87, 89, 91, 94, 98; teeth 73–74, 125

Taung, South Africa 156, 157, 159, 162, 246; distribution of large, medium, and small animals **157**; fossil baboon skull **160**; small animal fossils 141, 157

Tobias, Phillip Valentine 50–52, 61–63, 67, 82, 93–94, 108–109, 112, 141, 146–147, 184, 224, 227, **236**; dispute with Lee Berger 188–190; friction with Ron Clarke 228–256; papers by 193, 205, 231; personality 64, 140; quoted 69, 100; retirement 213, 226

Tools: bifaces 40, 41, 289; bone 151, 268, 269, 270, 290; chimpanzees 137; Olduwan 40, 239; stone 8–10, 37, 40–42, 92, 134, 151, 225, 239, 268, 270, 289–290, 292

Transvaal Museum, Pretoria, South Africa 85, 91, 97, 146, 151, 228, 246, 274

Turkana Boy 190; classified as *Homo erectus* 117, 118; damage to skull 123–124; discovery 39, 117; physique 39–40, 117–118; skeleton 124, 197, 274

Turkana, Lake, Kenya 54, 55, 111, 117; fossil fields 32–33, 38–39, 57, 191–192, 197

U

University of the Witwatersrand, Johannesburg, South Africa: hominid vault 242–243; Taung skull 89, 228

V

Volcanoes 104, 110, 115–116, 131

W

Walker, Alan 39, 118, 123–124, 200, 275

White, Tim 32, 36–37, 46, 49–50, 54, 64, 67, 114, 118, 164, 169–170, 190, 201–209, 211, 222, 245, 305; fossil discovery 119; Laetoli expedition 116; paper by 47; personality 204–205, 214, 215

Witwatersrand (region), South Africa 14, 71, 127, 141, 261, 284

Z

Zambia: Broken Hill Man 43, 287–288

Zinjanthropus boisei 63, 107–108, 110

PHOTO CREDITS

BIBLIOGRAPHY AND RELATED READINGS

Berger, L.R., (1992) "Early hominid fossils discovered at Gladysvale Cave, South Africa." *South African Journal of Science* 88, 362.

Berger, L.R., (1993) "A preliminary estimate of the age of the Gladysvale australopithecine site." *Palaeontologea Africa* 30, 51-55.

Berger, L.R., Keyser, A.W., & Tobias, P.V., (1993) "Gladysvale: first early hominid site discovered in South Africa since 1948." *America Journal of Physical Anthropology* 92, 107-111.

Berger, L.R., Menter, C.G. & Thackeray, J.F., (1994) "The renewal of excavation activities at Kromdraai, South Africa." *South African Journal of Science* 90, 223-226.

Berger, L.R. & Clarke, R.J., (1995) "Bird of prey involvement in the collection of the Taung child fauna." *Journal of Human Evolution* 29 (3), 275-299.

Berger, L.R., Pickford, M. & Thackeray, J.F., (1995) "A Plio-Pleistocene hominid upper central incisor from the Cooper's site, South Africa." *South African Journal of Science* 91, 541-542.

Berger, L.R. & Parkington, J.P., (1995) "A new Pleistocene hominid bearing locality at Hoedjiespunt, South Africa." *American Journal of Physical Anthropology* 96, 601-609.

Berger, L.R., (1996) "Birds of prey and the death of the Taung child." *South African Journal of Science* 92, 62.

Berger, L.R., (1996) *Is Africa the cradle of humankind? Nature, God and Humanity* (ed. C. du Toit) UNISA press, Pretoria, pp 226-243.

Berger, L.R. & Tobias, P.V., (1996) "A chimpanzee-like tibia from Sterkfontein, South Africa and its implications for interpretations of bipedalism in Australopithecus africanus." *Journal of Human Evolution.*

Berger, L.R. & Clarke, R.J., (1996) "Lifting the Taung child." *Nature*, 378, 670.

Berger, L.R. & Brink, J., (1996) "Middle Pleistocene fossil fauna including a human patella from the Riet River, South Africa." *South African Journal Science* 92, 277-278.

Broom, R. & Scheepers G.W.H., (1946) *The South African Fossil Ape-Men: The Australopithecines.* Transvaal Museum, Pretoria.

Brown, A.C., (1977) *A History of Scientific Endeavour in South Africa.* Royal Society of South Africa, University of Cape Town.

Brown, J., (1995) *Charles Darwin, Voyaging.* Jonathan Cape London.

Caird, R., (1994) *Ape Man – The Story of Human Evolution.* Boxtree Publishing, London.

Cameron, N., Berger, L.R. & McKee, J.K., (1994) "Training for Africans in Africa." *Nature*, 372, 589.

Cluver, M.A., (1978) "Fossil Reptiles of the South African Karoo." South African Museum, Cape Town.

Dart, R., (1957) "The Osteodontrokeratic Culture of Australopithe cus Prometheus." Transvaal Museum, Pretoria.

Hendey, Q.B., (1982) "Langebaanweg – A Record of Past Life." South African Museum, Cape Town.

Johanson D.C. & Edey, M.A., (1981) *Lucy – the Beginnings of Humankind.* Granada Publishing.

Johanson, D.C. & Shreeve, J., (1989) *Lucy's Child – the Discovery of a Human Ancestor.* Penguin, William Morrow and Co, New York.

Johanson D.C. & Edgar, B., (1996) *From Lucy to Language.* Nevramont Publishing Company.

Jones S, Martin R & Pilbeam D., (1992) *The Cambridge Encyclopedia of Human Evolution.* Cambridge University Press.

Leakey, L.S.B., (1974) *By the Evidence, Memoirs 1932 – 1951.* Harcourt, Brace Jovanovich Inc, New York.

Leakey, M., (1984) *Disclosing the Past.* Weidenfeld and Nicolson, London.

Leakey, R. & Lewin, R., (1992) *Origins Reconsidered – in Search of What Makes Us Human.* Little, Brown and Co, London.

Leakey, R., (1994) "The Origin of Humankind." *Science Masters.* Harper Collins, New York.

Leakey, R. & Lewin, R., (1996) *The Sixth Extinction – Biodiversity and its Survival.* Weidenfeld and Nicolson, London.

Lewis-Williams D. & Dowson T., (1989) *Images of Power–Understanding Bushman Rock Art.* Southern Book Publishers, Johannesburg.

Mason, R., (1969) *Prehistory of the Transvaal.* Witwatersrand University Press, Johannesburg.

McHenry, H.M. & Berger, L.R., (1998) "Limb proportions in Australopithecus africanus and the origins of the genus Homo." *Journal of Human Evolution* 35, 1-22.

McHenry, H.M. & Berger, L.R., (1998) "Limb lengths in Australopithecus africanus and the origins of the genus Homo." *South African Journal of Science* 94, 447-450.

McKee, J.K., Thackeray, J.F. & Berger, L.R., (1995) "Faunal assemblage seriation of South African Pliocene and Pleistocene fossil deposits." *America Journal of Physical Anthropology* 96, 235-250.

Roberts, D. & Berger, L.R., (1997) "117k year old human footprints from Langebaan Laggon, South Africa." *South African Journal of Science* 93, 349-350.

Schmid, P. & Berger, L.R., (1997) "A hominiod phalanx from the Gladysvale site, South Africa." *South African Journal of Science.*

Shreeve, J., (1995) *The Neanderthal Enigma – Solving the Mystery of Modern Human Origins.* William Morrow and Co, Inc, New York.

Stringer, C. & McKie R., (1996) *African Exodus – the Origins of Modern Humanity.* Jonathan Cape, London.

Tattersall, I., (1995) *The Fossil Trail.* Oxford University Press.

Taylor, T., (1996) *A Prehistory of Sex – Four Million Years of Human Sexual Culture.* Fourth Estate, London.

Terry, R., (1974) "Raymond Dart, Taung 50th Anniversary Commemoration Booklet." Museum of Man and Science, Johannesburg.

Tobias, P.V., (1984) *Dart, Taung and the "Missing Link."* Witwatersrand University Press, Johannesburg.

Walker, A., & Shipman, P., (1996) *The Wisdom of Bones – in Search of Human Origins*. Weidenfeld and Nicolson, London.

Woodhouse, H.C., (1971) *Archaeology in Southern Africa*. Purnell and Sons, Cape Town.

National Geographic Magazine

Dawn of Humans Series:

September 1995: The Farthest Horizon: Maeve Leakey

March 1996: Face to Face with Lucy's Family: Don Johanson

July 1997: The First Europeans: Rick Gore

September 1997: Tracking the First of Our Kind: Rick Gore

August 1998: Redrawing Our Family Tree: Lee Berger

Also:

March 1992: Apes and Humans: Eugene Linden

December 1995: Jane Goodall: Peter Miller

ACKNOWLEDGMENTS

A book of this nature is not written by just two people but is in fact the collaborative efforts of dozens of scientific colleagues, students, and technical staff that form the scientific organization that makes it all happen. To these people, too numerous to mention individually, we are eternally grateful. They know who they are.

Our thanks to our employers the University of the Witwatersrand and Primedia and Metropolis for allowing us the time and space to pursue this project.

A special thanks to the funders of the research endeavors in South Africa, in particular the Trustees of the Palaeo-Anthropological Scientific Trust (PAST) Tony Trahar, Steve Anderson, John Cruise, Clifford Elphik, Nick Frangos, Paul Gibbs, John Hunt, Malcolm McCulloch, Johnny Mosendane, John Nash, Mark Read, Norman Segal, Mary Slack, John Vorster, and the operational manager Christine Read. They saved the science here in the early nineties. Without these individuals' tireless efforts to provide funds for the heavy costs of field work there would be no discoveries. To our other major funders of research, the National Geographic Society, the Standard Bank of South Africa, the Anglo American Corporation, and the University of the Witwatersrand's Research Office, we offer our thanks and hope that the future of paleoanthropology in South Africa is as bright as its past.

To the late Gavin Relly, founding Chairman of PAST, we only wish he was here now to see the fruits of his endeavors.

A personal thank you to Harry Oppenheimer for his long-term friendship and individual support.

Our thanks to the *Sunday Times*, Independent Newspapers (in particular the *Star*), and the *Mail* and *Guardian* for the use of their archives.

To Owen Laster at the William Morris Agency, a genuinely nice man; his wisdom is inspiring.

To Kevin Mulroy and the rest of the editorial staff at National Geographic Society. Amazing is the only word.

Brett Hilton-Barber would like to thank the Hilton-Barber family of Tana, Dave, Steven, and Bridget for their support; Pauli and Murray Grindrod for providing the most conducive environment at Hermanus and Nottingham Road for writing this book; Paul and Michelle Jenkins for advice and encouragement; and most of all, Josie Hilton-Barber, whose love, commitment, and critical concern made this all possible.

I would like to thank my extended family in both Africa and the United States, especially my mother and father for always being there and encouraging my passion. The Smilg family in South Africa deserve a special thank you for allowing their daughter to run off with a fossil finder. To my wife, Jackie, my daughter, Megan, and son, Matthew, I cherish your love and support. You all make my life rich.